AI-Powered Commerce

Building the products and services of the future with Commerce.AI

Andy Pandharikar

Frederik Bussler

Packt>

BIRMINGHAM—MUMBAI

AI-Powered Commerce

Group Product Managers: Gebin George

Publishing Product Manager: Ali Abidi

Senior Editors: Storm Mann and Nisha Cleetus

Content Development Editor: Nithya Sadanandan

Technical Editor: Karan Solanki

Copy Editor: Safis Editing

Project Coordinator: Ajesh Devavaram

Proofreader: Safis Editing

Indexer: Manju Arasan

Production Designer: Shyam Sundar Korumilli

First published: November 2021

Production reference: 1191121

Published by Packt Publishing Ltd.
Livery Place
35 Livery Street
Birmingham
B3 2PB, UK.

ISBN 978-1-80324-898-1

www.packt.com

To our hard-working team, innovative customers, and supportive investors.

– Andy Pandharikar

To the reader, may you have as much fun reading this as I had writing it.

– Frederik Bussler

Contributors

About the authors

Andy Pandharikar is the CEO and founder of Commerce.AI. His previous start-up was acquired by the Flipkart group, which later was acquired by Walmart for $16B. Prior to that, Andy held various product and engineering positions at Cisco. He has an M.S. in management science and engineering from Stanford University and has an executive degree from Harvard Business School.

Frederik Bussler is a consultant and analyst, with experience across innovative AI platforms such as Commerce.AI, Obviously.AI, and Apteo, as well as investment offices such as Supercap Digital, Maven 11 Capital, and Invictus Capital. He has featured in Forbes, Yahoo, and other outlets, and has presented for audiences including IBM and Nikkei.

About the reviewer

Joey Bertschler is a brand and marketing expert who has worked with some of the world's most influential companies and thought leaders. His work is featured in Forbes, Entrepreneur, Hacker Noon, and many other outlets.

He has shared his journey in front of thousands at leading events such as Step Conference Dubai. Mr. Bertschler has worked internationally, from AI firms operating in Japan and India to pioneering sustainability ventures in Nigeria.

Table of Contents

3

Understanding How to Predict Industry-Wide Trends Using Big Data

Section 2: How Top Brands Use Artificial Intelligence

4

Applying AI for Innovation – Luxury Goods Deep Dive

5

Applying AI for Innovation – Wireless Networking Deep Dive

6

Applying AI for Innovation –Consumer Electronics Deep Dive

7

Applying AI for Innovation – Restaurants Deep Dive

8

Applying AI for Innovation – Consumer Goods Deep Dive

Section 3: How to Use Commerce.AI for Product Ideation, Trend Analysis, and Predictions

9

Delivering Insights with Product AI

10

Delivering Insights with Service AI

11
Delivering Insights with Market AI

12
Delivering Insights with Voice Surveys

Preface

The future of commerce is here. AI-powered commerce is enabling businesses to offer new experiences and value to consumers. AI will play a critical role in the future of product and service innovation, transforming how brands engage with customers across multiple touchpoints.

Commerce.AI introduces you to the latest advances in AI and how it's being used to power new products and services across a variety of industries. You will learn about the latest advancements in AI, including deep learning, **generative adversarial networks (GANs)**, **natural language processing (NLP)**, and computer vision.

The book begins with an overview of different applications of AI for product and service innovation, including market opportunity identification, creating product ideas, and industry trend forecasting. You will then explore AI for innovation use cases across a number of industries, from consumer electronics to luxury goods. Finally, you'll learn how to use Commerce.AI's core features to empower your product and service teams to create innovative products and services that meet the needs of your customers.

Who this book is for

This book will guide you through the process of product and service innovation, no matter your pre-existing skillset. Whether you're an AI developer, a product manager, an analyst, or a consumer insights professional, this book will teach you everything you need to use the power of AI for innovation.

What this book covers

Chapter 1, *Improving Market Opportunity Identification*, focuses on using AI for market opportunity identification and how it can be used to better understand customers' interests in addition to their needs and desires. This is an important step for any company looking to refine its value proposition, modify customer experience, or create new products or services.

Chapter 2, *Creating Product Ideas*, explains how to leverage AI and machine learning to find product ideas. We cover the concepts of NLP and large language models to generate product ideas. We explain how you can use AI to analyze market data to create relevant and in-demand product ideas.

Chapter 3, Understanding How to Predict Industry-Wide Trends Using Big Data, explains how to use advanced machine learning and data science techniques to predict future trends. It explains what big data is, how it's structured, and how you can leverage it to see into the future. This is vital to creating successful products or services because it helps you understand what consumers will want in the future.

Chapter 4, Applying AI for Innovation – Luxury Goods Deep Dive, explores how AI is used for innovation in the luxury goods industry. It explores how luxury goods companies are using this technology to create personalization, improve customer experience, and develop new products.

Chapter 5, Applying AI for Innovation – Wireless Networking Deep Dive, looks at how AI is used in the wireless networking industry. It explores how AI is applied to optimize KPIs such as star ratings, bestseller rankings, product sentiment, and conversion.

Chapter 6, Applying AI for Innovation – Consumer Electronics Deep Dive, analyzes how AI is used for innovation in the consumer electronics industry. It explores how this technology is applied to understand and optimize product positioning, brand research, idea generation, insights extraction, and more.

Chapter 7, Applying AI for Innovation – Restaurants Deep Dive, explains how data is a critical tool for restaurant innovation teams. You'll learn how to use data and AI to inform your restaurant strategy.

Chapter 8, Applying AI for Innovation – Consumer Goods Deep Dive, lays out how AI is being utilized in the consumer goods industry. It examines how AI can be used for consumer goods market intelligence, generating content, analyzing sentiment, and so much more.

Chapter 9, Delivering Insights with Product AI, is an exploration of Commerce.AI's Product AI features, which empower product innovation teams to research, develop, launch, and track winning products.

Chapter 10, Delivering Insights with Service AI, explores Commerce.AI's Service AI features, which enable service innovation teams to build next-generation experiences and reputation by truly understanding their customers and the rest of the market.

Chapter 11, Delivering Insights with Market AI, explores Commerce.AI's Market AI features, which enable teams to uncover facets, use cases, and topics to determine changing customer behavior, trending products, emerging brands, and new product opportunities.

Chapter 12, Delivering Insights with Voice Surveys, dives into Commerce.AI voice surveys, which use speech recognition and natural language understanding. You'll learn how to gain unparalleled insights into what consumers are thinking.

To get the most out of this book

You will need a stable Internet connection and a Python-compatible IDE, on either Windows, macOS, or Linux. All code examples have been tested using Google Colaboratory. However, they will work with offline IDEs and other environments as well.

Software/hardware covered in the book	Operating system requirements
Commerce.AI	Any web browser
Python pandas and other libraries	Windows, macOS, or Linux
GPT-J	Windows, macOS, or Linux
AI21 Studio	Windows, macOS, or Linux

> **Note**
>
> If you are using the digital version of this book, we advise you to type the code yourself or access the code from the book's GitHub repository (a link is available in the next section). Doing so will help you avoid any potential errors related to the copying and pasting of code.

During or after reading the book, we encourage you to sign up for a free trial of Commerce.AI at `https://www.commerce.ai/contact`.

Download the example code files

You can download the example code files for this book from GitHub at `https://github.com/PacktPublishing/AI-Powered-Commerce`. If there's an update to the code, it will be updated in the GitHub repository.

We also have other code bundles from our rich catalog of books and videos available at `https://github.com/PacktPublishing/`. Check them out!

Download the color images

We also provide a PDF file that has color images of the screenshots and diagrams used in this book. You can download it here: `https://static.packt-cdn.com/downloads/9781803248981_ColorImages.pdf`.

Conventions used

There are a number of text conventions used throughout this book.

`Code in text`: Indicates code words in text, database table names, folder names, filenames, file extensions, pathnames, dummy URLs, user input, and Twitter handles. Here is an example: "Take a look at more positive reviews, with a positive `polarity` and fairly low `subjectivity`."

A block of code is set as follows:

```
s = df['Reviews']
df['Reviews'] = df['Reviews'].astype(str)
df = df[df['Reviews'] == s]
df[['polarity', 'subjectivity']] = df['Reviews'].apply(lambda
    Text: pd.Series(TextBlob(Text).sentiment))
```

When we wish to draw your attention to a particular part of a code block, the relevant lines or items are set in bold:

```
s = df['Reviews']
df['Reviews'] = df['Reviews'].astype(str)
df = df[df['Reviews'] == s]
df[['polarity', 'subjectivity']] = df['Reviews'].apply(lambda
    Text: pd.Series(TextBlob(Text).sentiment))
```

Bold: Indicates a new term, an important word, or words that you see onscreen. For instance, words in menus or dialog boxes appear in **bold**. Here is an example: "Even positive reviews complain about the range, such as one review that says simply **Good short range**."

Tips or important notes
Appear like this.

Get in touch

Feedback from our readers is always welcome.

General feedback: If you have questions about any aspect of this book, email us at customercare@packtpub.com and mention the book title in the subject of your message.

Errata: Although we have taken every care to ensure the accuracy of our content, mistakes do happen. If you have found a mistake in this book, we would be grateful if you would report this to us. Please visit www.packtpub.com/support/errata and fill in the form.

Piracy: If you come across any illegal copies of our works in any form on the internet, we would be grateful if you would provide us with the location address or website name. Please contact us at copyright@packt.com with a link to the material.

If you are interested in becoming an author: If there is a topic that you have expertise in and you are interested in either writing or contributing to a book, please visit authors.packtpub.com.

Share Your Thoughts

Once you've read *AI-Powered Commerce*, we'd love to hear your thoughts! Scan the QR code below to go straight to the Amazon review page for this book and share your feedback.

https://packt.link/r/180324898X

Your review is important to us and the tech community and will help us make sure we're delivering excellent quality content.

Section 1: Benefits of AI-Powered Commerce

In the first section of this book, you will learn the benefits associated with AI-powered commerce and understand how to use AI systems for your own business.

This section comprises the following chapters:

- *Chapter 1, Improving Market Opportunity Identification*
- *Chapter 2, Creating Product Ideas*
- *Chapter 3, Understanding How to Predict Industry-Wide Trends Using Big Data*

1
Improving Market Opportunity Identification

Just a few years ago, every conversation about **artificial intelligence** (**AI**) seemed to end with an apocalyptic prediction. In 2014, Elon Musk said that, with AI, we are *summoning the demon*, while Stephen Hawking said that AI *could spell the end of the human race*. More recently, however, things have begun to change. AI has gone from being a scary black box to something people can use for a variety of use cases.

This shift is because these technologies are finally being explored at scale, including by product teams for market opportunity identification. AI hasn't always been used in the industry. In fact, it started out as a scientific curiosity. In the 1950s, computer scientist John McCarthy wanted to see whether it was possible to build machines that could learn how to do tasks such as play chess themselves. Today, AI is everywhere.

In this chapter, we'll explore how market opportunity identification can be improved with big data and AI, covering the following topics:

- Identifying market opportunities the traditional way
- Big data challenges in market opportunity identification
- Using AI for market opportunity identification
- Exploring AI-powered market reports

Market opportunity identification is important for a product team to identify an unmet need. It helps them find out how their product will stand out in the market and what they need to do in order to grow. They need to identify the competitive landscape, define the market opportunity, and use this to create a value proposition.

Furthermore, market opportunity identification sets the groundwork for later chapters, including creating product ideas and predicting future market trends.

Identifying market opportunities the traditional way

When it comes to commerce, venturing blindly into the unknown is a recipe for failure. Belief in a market opportunity is not enough – there needs to be hard data. Market opportunity identification is the process of acquiring and analyzing data, from any source, to understand the potential size of a market and the potential share of that market that you can capture.

Broadly speaking, the market should be considered in three segments:

- **External**: External markets are those that are already established and they may be served by other companies.
- **Internal**: Internal markets are those that exist within the company but they may not be recognized as such yet.
- **Potential**: Potential markets are those that have not been identified yet.

Traditional market opportunity identification can be done in a number of ways:

- Firstly, it can be done by **surveying the consumers**. Market surveys are one of the most effective ways to find out what customers want and how much they are willing to spend for it. This not only gives a company an idea of what to produce but also helps figure out how much money they will make from their products.

- Secondly, market opportunity identification can be done through **brainstorming techniques** such as the *5 Whys* technique. The 5 Whys technique is simple and can be used as a brainstorming exercise by asking *why?* five times in response to a topic, problem, or issue.

- Thirdly, market opportunity identification can be done by **analyzing internal data** such as surveys and interviews that have already been done before. Often, large companies will have a lot of unorganized, unstructured information that is divided across departments and projects, from Google Surveys to JotForm to SurveyMonkey. When this data is analyzed, they may find opportunities they weren't aware of before.

- Finally, traditional market opportunity identification can be done by **examining external data** and data from social media platforms or competitors in the same industry. This includes sites such as LinkedIn or Facebook where both companies' and individuals' data can prove valuable.

One tool that is becoming very popular for market research is **Google Trends**, which allows people to examine the search volume in a particular area of interest over a certain period of time. For example, if you were interested in finding out about the popularity of rooftop gardens in Los Angeles, you could type into Google Trends *Los Angeles Rooftop Gardens*, and analyze search trends over time. Additionally, Google Trends lets you easily export this data for further analysis, whether it's to create visualizations or merge with additional data.

Once potential market opportunities have been identified, then the company should consider the decision to either pursue or ignore them. If the company chooses to pursue an external, internal, or potential market, then it needs to consider which approach is best.

For example, if an *internal* market is being considered for pursuit, it may be best for the company to create an incentive and offer it to its employees in order to get them involved with the new product.

If an *external* market is being considered for pursuit, it may be best to invest in advertising campaigns to make customers aware of the new product.

If a *potential* market is being considered for pursuit, it may be best for the company to invest in research and development in order to create a product that will appeal to this new market segment.

There is also a difference between **customer-driven identification** and **opportunity-driven identification**. Customer-driven identification is where an organization determines a marketing need based on what customers want or need, whereas opportunity-driven identification is where an organization identifies areas of potential value based on strengths and weaknesses.

For example, if an organization is strong in manufacturing but weak in marketing, then opportunities for enhancement may be found in the marketing area that would not have been found had they done customer-driven identification.

While there are a number of benefits to traditional market opportunity identification, these methods fall short in the modern world, which is faced with big data challenges.

Big data challenges in market opportunity identification

Big data has become a buzzword in the product community.

Big data involves the speed at which data is generated, the amount of data that's generated, the types of questions that can be answered with it, and the number of sources it's coming from. In short, big data is about more than just size.

Big data describes the veritable explosion of data we're seeing from billions of people accessing the internet.

Product teams want to use big data to identify new market opportunities and new ways to target their customers. Yet, many companies are struggling with how to collect and analyze big data, particularly when the data is scattered across different sources.

Business executives are asking questions such as: *How can we get useful information from all of this disparate data? How can we make better strategic decisions with our big data? How can we address the challenges of storing, organizing, and managing big data? And how can we increase the value of our big data?*

Commonly, there are a few major challenges associated with managing big data for market opportunity identification:

- The first challenge is that **it's difficult to find relevant predictive patterns in a sea of unrelated variables**. For example, using traditional **business intelligence (BI)** techniques such as data munging and manual data mining, it might take weeks or even months to discover a predictive pattern in a large database of customer survey results. Imagine that your dataset includes demographics, firmographics, psychographics, purchase data, reviews, and more. That's a lot of information to look through, and a lot of variables that might not seem relevant at first glance.

- A second challenge is that **traditional BI tools are not designed for efficient discovery of predictive patterns when analyzing big data**, which is increasing in volume at a faster pace than traditional databases are being updated. Not only does this make it difficult to keep up with the latest insights, but it also becomes more costly and time-consuming to build insights into existing systems.

- Another major challenge with big data is **simply getting it in the first place**.

Large companies such as Google and Amazon have access to tremendous amounts of computing power and virtually unlimited storage thanks to their substantial investments in hardware and software services, but for smaller organizations – even those that are eager to harness the potential of big data – the story is different.

Data is coming in faster than ever before. The explosion of data being generated has outpaced the capabilities of traditional database systems to keep up with growth, but setting up and maintaining systems such as AWS or GCP requires dedicated engineers, given the requirements for technical know-how on scaling, security, data pipelines, and more.

Therefore, the problem is often that businesses lack the necessary resources that would allow them to collect, aggregate, analyze, and interpret such volumes of information without investing large sums of money in server clusters or other specialized infrastructure.

With AI, companies can solve these challenges more easily, and analyze big data to improve market opportunity identification.

Using AI for market opportunity identification

The most important thing to remember about market opportunity identification with AI is that it is not about creating a *magic wand* that will instantly identify all of the major new market opportunities. That would be a data scientist's or software developer's dream. But in the world of marketing, where success depends on being fast, efficient, and smart, those are the characteristics of an outright nightmare.

Marketing and product development professionals are constantly under pressure to deliver new products and services to market – and be first to market with them. They need to do so in a cost-effective manner, without sacrificing quality, and often with limited resources.

AI offers speed, efficiency, and smartness. With models deployed on scalable servers, AI can scan a huge amount of data and identify patterns that humans wouldn't see. This means that millions of data points can be analyzed within hours or even minutes.

AI is being used in a number of verticals, such as autonomous vehicles, facial recognition, and fraud detection.

By using AI for market opportunity identification, marketers can free themselves from the burdens of manually reviewing and analyzing information on countless possible markets or new product categories. They can offload tasks such as data collection and the creation of new product ideas to machines.

The best way to use AI for market opportunity identification is by *focusing on real problems that need real solutions*. For example, marketers may want to analyze their existing customer base for potential new product categories they could enter into – or specific products they could create for those existing customers.

Another example is using AI to identify new geographic markets for existing brands – or even entirely new brands – by looking at various combinations of customer demographics, buying habits, lifestyle choices, and other criteria in each region of the world.

In either case, AI can assist marketers by helping them use data already in their possession more effectively than they could otherwise (by automating some data collection tasks and streamlining others). It can also help overcome some key barriers, including assessing whether any given potential market or product category is truly scalable.

An additional benefit of using AI for market opportunity identification is that you can also use it as a tool for demonstrating your commitment to innovation and differentiating your brand(s) from competitors' offerings – which can also boost your brand awareness, reputation, and overall value.

As we've explored, AI has a number of broad benefits for product teams. Let's dive into a specific use case: **AI-powered market reports**.

Exploring AI-powered market reports

AI has transformed **machine learning**, a branch of artificial intelligence that automates the identification of patterns in data. The big data analytics industry is abuzz with the possibilities of AI-generated market reports (`www.commerce.ai/reports`) in identifying opportunities and trends.

An AI engine can be used for a variety of purposes, including generating market reports to identify potential opportunities for improvement in marketing campaigns. The technology is ripe for marketing professionals to take advantage of, making it easier than ever to generate local or global reports using existing data.

Here are some of the benefits of using AI-generated **market reports**:

- **Automation**: Automating repetitive tasks such as generating reports can help marketing professionals save valuable time and energy.

- **Cost-effectiveness**: Generating report templates could cost less, with AI used in the background to quickly adapt the template for local or global needs.

- **Overall quality improvement**: By using AI, marketers could have more accurate, precise, and complete information to use in their campaign planning efforts.

- **Data tracking**: Using machine learning algorithms to analyze large amounts of data could help marketers gain valuable insights into consumer preferences and buying trends.

Highlighting these benefits is not meant to imply that there are no drawbacks to using AI-generated market reports. Instead, it is meant to highlight several *key benefits* that can help you make a final decision about whether or not you should use this technology in your business operations moving forward.

With a new AI-generated market reports feature, Commerce.AI now delivers high-quality insights straight to the public.

Previously available exclusively to those with access to Commerce.AI's data engine, market reports are now available to all.

Market reports analysis is based on consumer feedback and offers valuable insight into how people consume products and services. With the press of a button or a simple search, anyone can now access market reports across 10,000 categories.

Commerce.AI's AI-generated reports are a concrete way to find your next product idea or seek a new customer base:

Market Reports

Fast moving AI-extracted insights about competitors, trends and opportunities
in thousands of markets.

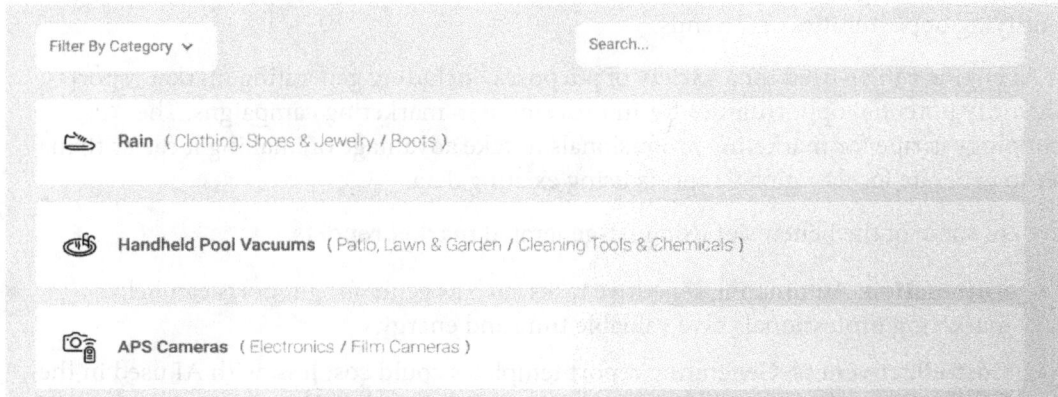

Filter By Category ⌄		Search...

🐾 **Rain** (Clothing, Shoes & Jewelry / Boots)

🦀 **Handheld Pool Vacuums** (Patio, Lawn & Garden / Cleaning Tools & Chemicals)

📷 **APS Cameras** (Electronics / Film Cameras)

Figure 1.1 – A sample of Commerce.AI's AI-generated market reports

For example, searching for *gaming keyboards* will show us a high opportunity score, as this niche market category has grown tremendously during pandemic-fueled lockdowns. As Commerce.AI continues to grow, the quality of its insights will only improve and become more accurate as more product data related to new trends in the marketplace is analyzed.

Commerce AI's data engine has delivered over 140 million dollars in revenue through insights across market categories. The market reports are an incredible resource built around billions of unstructured product data points, from sources such as forums and blogs, surveys, videos, customer support tickets, CRMs, and more. They provide a glimpse into how people consume products and services, with the potential to help you find your next big idea.

When it comes to innovation and product development, market research is invaluable. It provides insights into how people consume products and services, which can be used to inform the next idea that will take on the world. Without using consumer feedback as a starting point for conceptualizing original ideas, you're limiting possible endeavors.

Commerce.AI's market research reports include a summary for each category, the fastest-growing brands, the bestseller, top-rated products, the number of products and reviews, and more. All this data is consolidated into a single value – the opportunity meter – showing the size of the market opportunity at hand.

AI calculates this by taking into account the average market size, the number of competitors in each category, and the potential for growth, creating a *line of best fit* between the data points:

AI Generated Summary

Our data analytics platform provides a 360 view of all consumer markets. We monitor online channels and provide structured data to the industry. Our monitoring identified 1,986,480 datapoints for the Kabuki Brushes category. There was an opportunity level of 53 for a new brand or product in this sector. We detected increased sales in the Kabuki Brushes with an opportunity level of 53 with 288 products.

Fast Growing Brands

Daubigny, e.l.f., KESHIMA, EcoTools, Existing Beauty, TEXAMO, Coco & Eve

Bestseller

Large Flat Top Kabuki Foundation Brush By Keshima - Premium Makeup Brush for Liquid, Cream, and Powder - Buffing, Blending, and Face Brush, 1.6" Top Diameter **(4.6 stars, $10.00, 17,821 ratings)**

OPPORTUNITY METER

41-54%

BRANDS
7+

PRODUCTS
288+

REVIEWS
16,389+

Figure 1.2 – A sample AI-generated market report

Traditional market research is expensive, inaccessible, and confusing. Commerce.AI's new AI-generated market reports feature solves these problems as well as the issue of a limited time frame in exploring market opportunities.

Market reports are available without having to pay an exorbitant fee for access to data-driven consumer insights.

Commerce.AI's system relies on a variety of different types of product data points, including unstructured data, which makes up 95% of the web. These sources include Amazon, Walmart, Target, and other commerce sources that are crucial for product teams.

All of this, compiled together and analyzed with AI, produces marketplace insights providing valuable knowledge into how people consume products and services.

With Commerce.AI's new marketplace insights, you can be confident in the next product idea you're considering as well as your current markets. With data from 10,000 market categories scanned and reports available to all, there has never been this scale of access to high-quality insights.

Skip the expensive market research firms that don't have a clue about how people are spending money these days, and take advantage of the power of billions of data points analyzed for you with AI-generated market reports.

Of course, there are other outlets for market reports, from Gartner to Nielsen. However, without the power of AI, these traditional market reports can be subjected to biases, while missing out on analyzing billions of data points. Moreover, AI-generated reports are far more cost-effective, costing a fraction of their traditional counterparts.

For instance, Gartner's research subscription costs around $30,000 a year, since the overhead that goes into traditional reports is high. This is because it takes highly paid teams of analysts many hours to produce a report.

In contrast, AI-generated reports compile billions of data points across the web using **natural language processing** (**NLP**) software; as such, they take only minutes to generate and cost far less. NLP works by computationally analyzing patterns in language, which gets encoded as numerical token values. It's based on the idea that all language is structured around a core set of elements, such as words and phrases, and these can be combined in many different ways to express an enormous range of ideas in a vector space.

Billions of data points are boiled down into a single opportunity score for any given market, providing unparalleled accuracy and scale, without sacrificing clarity and ease of understanding. These data points provide organizations with unequaled insights on how to advance their strategy and prepare for expansion around the globe.

Summary

Clearly, AI has the potential to radically improve the quality and speed of market opportunity identification.

But it's not just about finding better ways to do what researchers have always done or even automating tasks. Instead, AI can help us see new opportunities that we never could have uncovered before.

In a traditional BI environment, analysts must sift through hours of data on thousands of companies in order to identify promising new targets.

With an AI solution built around unstructured data sources (such as text documents or images), this task becomes straightforward; an AI system can simply scan millions of documents for identifying keywords or patterns and then provide recommended matches with commercial significance.

Commerce.AI is one of many advanced technologies available to product teams. It uses a range of machine learning techniques to find insights based on structured and unstructured data.

Understanding market opportunity identification is important for product teams to lay the groundwork for the product development process. Traditional methods have been in place for generations, and are still relevant today, while AI-based methods can give teams a competitive advantage. This understanding is crucial for the upcoming chapters, as product development should be based on a deep understanding of the market. Moving on from this understanding, we'll dive deeper into topics such as building, selecting, and iterating product ideas.

In the next chapter, we'll explore how the product creation process can be improved with AI.

2
Creating Product Ideas

Coming up with great product ideas isn't easy. It's both an art and a science, and those with the ability to come up with great ideas are remembered in history, from people such as Steve Jobs to Mark Zuckerberg.

In this chapter, we'll explore the **five pillars of AI**, which are driving new and innovative ways to create product ideas: language understanding, visual understanding, information extraction, information organization, and creative AI. Understanding the pillars of AI will help you become more strategic about how you plan, manage, and invest in your AI projects.

For instance, some product teams might like to use AI to generate new product designs, which would focus on the pillar of *visual understanding*, while others might like to use AI to understand desired customer features that will lay the framework for new products, which would involve pillars such as *language understanding* and *information extraction*.

After covering that, we'll cover building, selecting, and iterating product ideas. Finally, you'll learn how to use Commerce.AI to improve the product ideation process and take advantage of billions of data points to gain a competitive advantage.

In this chapter, we will cover the following topics:

- Understanding the pillars of AI
- Why is product ideation so hard?
- Using Commerce.AI for creative AI
- Building product ideas
- Selecting product ideas
- Iterating product ideas

Product ideation is crucial to business success. In the past, companies have failed to create products customers wanted because they didn't know what the customer wanted until *after* the product was built. But that approach is now increasingly obsolete. The internet has given companies unprecedented access to customer data and insights about their customers at any given time, allowing them to build better products faster than ever before, aided by the power of AI.

Understanding the pillars of AI

These are the five pillars of AI, which lay the groundwork for using AI for product innovation:

1. **Language understanding**
2. **Visual understanding**
3. **Information extraction**
4. **Information organization**
5. **Creative AI**

When you combine the first four pillars with creativity, you get what's called *creative AI*. In other words, the first four pillars are needed to create the data structure that fuels a gamut of creative use cases.

Creative AI is an advanced form of artificial intelligence that can solve problems previously thought impossible for machines, whether that's designing wholly new products or coming up with truly innovative ideas, such as how Google used AI to design *rounded and organic* computer chips much faster than human engineers, or how designers at Autodesk use AI to design skeletal scaffolds that are lighter, stronger, and more efficient than regular designs. In this section, we'll explore the five pillars of AI, and how they tie into product creation, in greater detail.

Language understanding

Language understanding is the ability to read users' minds.

One of the most important pillars of AI innovation is language understanding, which allows machines to interpret human text and reasoning, and then return a response that the user can understand. The ability to do this is often referred to as **natural language processing** (**NLP**). NLP has been around for decades, but just recently it has witnessed significant advances through deep learning algorithms.

In contrast to traditional machine learning methods, deep learning can identify patterns in data using neural networks. This approach makes use of large amounts of datasets and produces accurate results at higher speeds than previous methods. Deep learning can not only predict future outcomes but also perceive the user's intent or state of mind from their voice or writing.

For example, if you ask a Google Home device about the weather in San Francisco tomorrow, it might respond with *It will be sunny with a high of 82 degrees Fahrenheit*. This is due to deep learning technology interpreting your spoken words as requests for information, and providing you with what it believes you want based on its vast knowledge base.

The point here is that today's modern NLP technology enables machines to understand our intentions better than ever before, which means we can effectively build conversational agents easier for a variety of tasks (for example, when scheduling a meeting). The good news is that because modern NLP technology isn't very complex, compared to other AI technologies such as image recognition, we don't have any shortage of examples where it could benefit our lives.

One way language understanding helps product teams is by generating new product ideas from existing ones. For example, say you have an existing product idea for a new hard hat for construction workers. You could use language understanding to automatically extend that idea into a smart hard hat that monitors the worker's location and warns them about their proximity to dangerous objects. At a high level, this is like autocomplete on steroids. We'll explore how this works in the *Transformers* section.

Another application of language understanding is helping companies scan online user reviews for specific features that users want to be added to their products. For example, if you go to Amazon right now and look at the reviews section for any popular product, you will see dozens of customer comments asking about the product features, either explicitly or implicitly. A popular snack includes comments such as *flavor is mediocre* and *the snacks are clumped together*.

With language understanding, these comments can be extracted and turned into a customer wishlist. Product teams can use these wishlists as feedback signals, telling them what their customers want from their products so that they can improve them over time.

Given that 95% of the data on the internet is unstructured and largely textual data, language understanding is a crucial pillar of AI. This technology applies to voice data as well, which enables new ways to collect and analyze customer feedback for market research, such as voice surveys.

In fact, at Commerce.AI, we've found that 95% of research participants prefer voice surveys over traditional survey forms.

Visual understanding

Visual understanding is about recognizing objects using images. An intelligent computer program can understand the elements of a picture and use that understanding to generate new ideas, and AI programs can also be trained on user feedback to develop new products based on what customers want. These programs can gather customer feedback, such as through our product data engine, voice surveys, or focus groups, and then turn it into product ideas using AI. This process produces an enormous amount of data about users' interests.

Let's explore how visual understanding is used in three different types of product innovation:

1. **Generating better insights on products and product feedback**

   First off, visual understanding addresses the task of finding data about people's needs, where an AI system can be trained to recognize objects within images. This enables better insights on products and product feedback.

   For instance, there are millions upon millions of products listed on Amazon. Factoring in Amazon Marketplace; the number is estimated to be around 350 million products. Many of these listings have not diligently listed all the visual information found within the photos. In other words, Amazon product photos are another crucial source of product data. Visual understanding can be used to understand these photos and their details, which is particularly useful when the textual descriptions fall short.

Apple uses facial recognition, a form of visual understanding, to enable users to effortlessly unlock their iPhones. They also use augmented reality, another type of visual understanding, to create personalized emojis.

2. **Developing new products by mining user feedback**

Beyond generating better insights, visual understanding can also help you develop *new* products and features. For example, many online reviews include images of the product while focusing on standout features, but also defects. Visual understanding can be used to analyze these images, in addition to regular text data, to help inform product teams of what products and features to change, and to avoid similar defects in the future.

3. **Developing solutions for users who are visually impaired or blind**

Let's look at one more benefit of visual understanding related to user experience innovation: Visual understanding is also used to improve software that allows blind computer users to access websites more easily. For example, many images on product listings lack *alt-text*, or the invisible description of images that are read aloud to blind users. With visual understanding, this alt-text can be generated automatically.

This can be done by using AI libraries such as OpenCV, which is a popular framework for computer vision and machine learning. OpenCV uses techniques such as **Convolutional Neural Networks (CNN)** to perform image classification tasks by extracting patterns and features to classify images based on what has been learned previously.

These neural networks might extract features such as color histograms, edges, and shapes in the image, or any other feature that makes it easier to identify and distinguish between different objects that, once identified, can be added as alt-text.

Information extraction

Information extraction is the process of extracting information from unstructured textual sources to enable finding entities, as well as classifying and storing them in a database. This is a big part of data science and AI.

At the time of writing, we can use a variety of tools to extract information from unstructured textual sources, such as NLP and deep learning.

By using information extraction on customer data, you can start to turn data into insights.

Using customer data

Information extraction is, of course, a huge part of extracting product data, but it's also used in the context of customer support services and can help you with questions such as the following:

- *What do customers say about our products?*

- *How do customers react to different claims in our product descriptions?*

- *Which words are most commonly mentioned in our product reviews?*

This customer data can come from a huge range of sources, including forums and blogs, surveys, videos, customer support tickets, CRMs, and more. The following screenshot highlights the very tip of the iceberg of customer data sources:

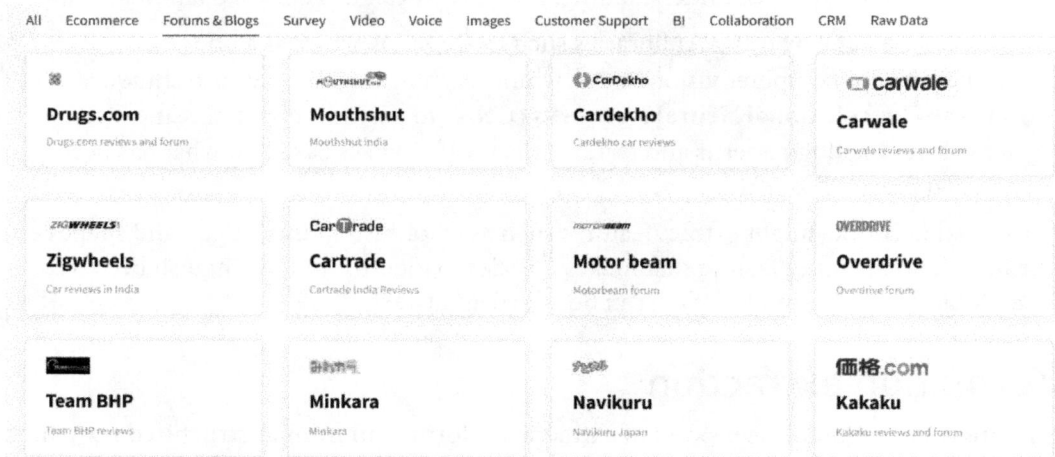

| All | Ecommerce | Forums & Blogs | Survey | Video | Voice | Images | Customer Support | BI | Collaboration | CRM | Raw Data |

Drugs.com
Drugs.com reviews and forum

Mouthshut
Mouthshut India

Cardekho
Cardekho car reviews

Carwale
Carwale reviews and forum

Zigwheels
Car reviews in India

Cartrade
Cartrade India Reviews

Motor beam
Motorbeam forum

Overdrive
Overdrive forum

Team BHP
Team BHP reviews

Minkara
Minkara

Navikuru
Navikuru Japan

Kakaku
Kakaku reviews and forum

Figure 2.1 – A sample of customer data sources

The holy grail is to answer customer questions preemptively or make suggestions based on what users have said so far. This kind of service might sound futuristic, but there are some companies already offering it. For example, Expedia has just released an AI bot that uses machine learning and NLP to help people book hotels. More specifically, this uses what's called **intent extraction**, which means finding out what kind of information a user wants. For example, if a user types `my card isn't working` in the chatbot, the system needs to process that as a request for payment information.

The intent extractor looks at all the possible intents and tries to match them with what the user has typed. At a technical level, this works by using a model trained on past conversations that were labeled with user intent. The model will understand patterns in the user language to match the user's messages with intents, even if the keywords aren't identical.

However, the biggest challenge here is figuring out how to scale up NLP for commercial purposes without getting trapped into using canned responses or preselected question sets. Chatbots should be able to have natural conversations with humans, and canned responses can stunt the human-like qualities of communication, which means that high-quality NLP tools are needed.

Using AI

This is where AI comes in handy: to generate new ideas that can be tested out before spending money on building a new chatbot or hiring new employees for customer service jobs.

While extraction is not the solution in itself, it will enable us to ask better questions about problems and identify opportunities for improvement. Data extraction provides a means to quickly learn about product reviews, sentiment, and market data. By providing insights into customer behavior (for example, what features users use most frequently and how often they encounter particular problems), we can ask better questions about why users face the problems they do, and how to fix them.

After extracting data, we can move on to organizing it.

Information organization

Information organization is the process of organizing extracted data to come up with greater insights.

Information organization creates structure and establishes the relationships between products, brands, and attributes, enabling the creation of new ideas, descriptions, and even ad copy. For example, the following screenshot highlights information about watch brands organized by Commerce.AI into a **Feed Summary**. From just a glance at this organized information, we can see that customers appreciate Casio's prices, but complain that they're made out of plastic and have bad batteries.

Feed Summary

>> **Top Brand**
Casio holds the largest review marketshare at **13.09%**.

>> **Top Brand Strengths**
Consumers praise **Casio** for **watch,great,price**.

>> **Top Brand Weaknesses**
Consumers complain about **Casio** for **plastic,dead_battery,returned**.

>> **Losing Brand**
BUREI is losing the most review marketshare at **-136** reviews (**-55.06%**).

>> **Outlier Brand**
Invicta review marketshare is growing the fastest at **540** reviews (**105.26%**).

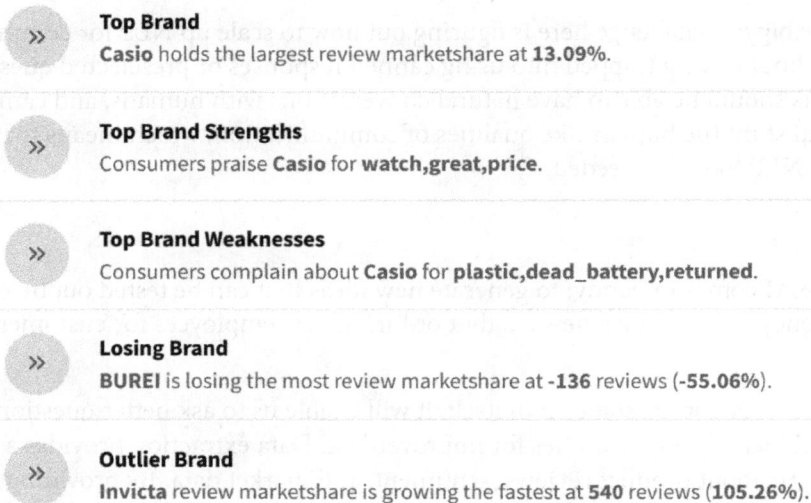

Figure 2.2 – A sample of organized information about watch brands

AI can organize information faster, cheaper, and more effectively than human beings. Let's look at five specific ways in which AI can help with information organization.

Reducing information gathering time

While collecting data is crucial for product innovation, it is also time-consuming. This is one of the biggest drawbacks of collecting data from users and customers.

More specifically, you need to collect massive amounts of user feedback or customer preferences over a certain period (which takes up a lot of the budget). But with the advent of powerful AI tools such as machine learning and NLP, you don't have to spend as many resources on data analysis anymore.

Instead, you can save tons of money by outsourcing data analysis to AI that knows how to read user feedback. This approach allows you not only to save resources but also to make more informed decisions about products and improve your business results faster.

Information organization depends on gathering the right product information across brands, product categories, and attributes. By using AI to speed up information gathering, we can make the right product data available for information organization, faster.

Eliminating duplicative data gathering

Another major drawback in running surveys or interviews is getting redundant responses from customers or users.

For example, if I want to find out about my customer preferences on various products available in the market today, I would ask various questions, such as, *Which smartphone do you currently own? What brand do you prefer? Are there any other features that come standard with these phones?*

With the help of artificial intelligence tools such as machine learning and NLP, you can get rid of repetitive feedback by automating tasks such as sending out surveys or conducting interviews automatically using chatbots instead.

Making better suggestions based on big data

There are times when we tend to rely on our judgment rather than analyzing what people want or need. Some examples of this are when a new product idea comes to mind, when we imagine a feature we'd like our products to have, and when we think about improving an aspect identified as an issue.

Since users often explicitly mention desired features and product attributes, whether it's in Amazon reviews or YouTube unboxing videos, it's a better idea to tap into this big data rather than relying on gut feeling. High-quality organized information requires big data, especially when it comes to understanding customers at scale. By using AI for big data analysis, companies can mine their data to find trends and patterns that they might not have been able to see before, enabling higher-quality information.

Gaining value from existing data

The most common method of feedback gathering that's used by companies today is asking users questions.

However, this feedback often already exists in the form of millions of online product reviews, both by the firm in question and its competitors. The problem with this data is that it's unstructured and unorganized, which means that there's untapped value. With AI, companies can finally structure and gain value from this existing data. As AI becomes more and more accessible, product teams are starting to take advantage of this unstructured data, but it's mostly still trapped under the technical burden of analyzing large amounts of raw data. As a result, product teams can still gain a competitive advantage by using AI.

Creating heatmaps

Machine learning helps generate heatmaps visualizing information gathered from online forms, filled up across multiple platforms such as web pages – both mobile websites and desktop sites – provided by services such as Google Analytics. This organized data helps companies before and during creative ideation. By organizing what users care about into a visual hierarchy, product teams can focus their creative ideation process on what truly matters. For example, suppose you're an automotive firm, and users are far more interested in your self-driving features than fast charging — this would inform your creative ideation process to focus on improving your self-driving features even further.

Creative AI

By using AI to generate ideas, we can build more products that delight customers and turn them into lifelong users. In a nutshell, we leverage the latest AI technologies, particularly large language models, to guide product ideation.

This process is called **concept generation with language models (CGLM)**. CGLM is an effective strategy for generating a range of novel concepts that meet user needs and desires.

Beyond generating natural language, creative AI can be used to design new products using a technology called **Generative Adversarial Networks (GANs)**. The idea of GANs is to train two competitive neural networks, where the first network generates fake images and the second network *discriminates* whether the images are real or fake. The generative model iteratively tries to fool the discriminative model until real images are indistinguishable from the fake ones. GANs were infamously applied to create *deepfake* photos and videos of celebrities and politicians, but they can also be used to generate product concepts, whether it's a new apartment layout or a sneaker design.

These types of creative AI support the product ideation process, which is typically a long, hard endeavor.

Why is product ideation so hard?

One of the most popular methods for generating new product ideas is through the use of brainstorming exercises, such as ideation exercises or mind mapping techniques, which are used by designers, architects, and engineers who are stuck or have reached dead ends in their creative processes. These methods are very useful because they enable people from different fields to come together and share different perspectives.

However, this method has limitations in terms of generating truly novel commercial products, given its focus on imagination rather than on actual customer needs and desires.

Another technique that's used by many companies today involves including customers in ideation processes, via surveys or focus groups, where they share their opinions about what they want. While this approach can be better for incorporating customer feedback, it's also costly and time-consuming. Domain experts have to manually take notes, collate and analyze feedback, and merge this with external data to move forward. They then need to find a way to collaborate with the actual product development teams to ensure that feedback is understood and used.

This lengthy, expensive process is also highly limited in terms of the data that's collected. A focus group can only include so many users. With AI, all available product data can be analyzed, including textual product reviews, product descriptions, video reviews, voice surveys, and more. AI is also much faster than humans, which means this data can be extracted, organized, and analyzed in real time, providing insights directly to product teams. This organized data acts as the fuel for creative AI as well, including product ideation. At the end of the day, product ideation is vital for success, lest your business end up as a graveyard firm that failed to innovate.

And it doesn't have to be burdensome, as Commerce.AI shows. Commerce.AI is a platform that allows you to easily build and manage your product ideas, from ideation through development and launch.

Using Commerce.AI for creative AI

As we've established, creative AI is powered by language understanding, visual understanding, information extraction, and information organization. With Commerce. AI, these components come together to let you generate new product ideas at will.

Commerce.AI's standard analytics include a product leaderboard, graphs of the top products and brands by market share over time, a review breakdown of the top products and brands (by stars), a market landscape graph, and more. The creative AI component uses this data, in addition to selected customer wishlist information, to generate new product ideas. Similarly, you can generate ad copy as well, which is typically a tedious, manual task, that can now be done effortlessly.

As we can see, AI can be creative, and a powerful tool for product teams to expedite their innovation process. Now that we have a background in AI and product ideation, let's use this knowledge to learn how to build, select, and iterate upon product ideas. Naturally, building product ideas is the first step, but not every idea will be a hit, which is why it's important to diligently select and iterate upon the best ideas.

Building product ideas

Innovating new product ideas is a serious challenge. With AI and, in particular, **large language models** (**LLMs**), product ideation becomes effortless.

LLMs can take massive amounts of textual information (whether it's text reviews, YouTube videos, or voice surveys) and generate novel text. The more data that has been annotated for the model, the better it can generate original ideas. For example, with Commerce.AI's product idea generator, machine learning is used to scan large amounts of product reviews from a given category, such as men's wristwatches, and then extract customer wishlist information.

The next step is taking this customer wishlist information and feeding it through an LLM to generate a new product idea. In the following example, given the wishlist points of `I just wish the hands glowed` and `I wish it had a metal caseback`, the LLM generates the following text:

```
A carbon steel watch with a glowing face. Under each hour, you
have a number of dots that light up in sequence from left to
right, then back to the left again when it reaches 12.
```

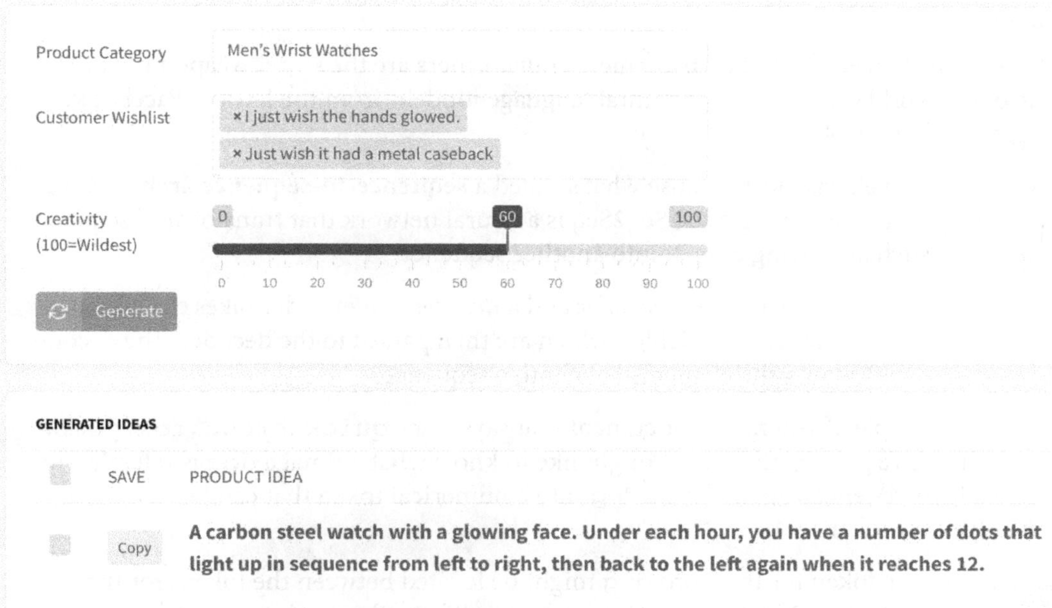

Product Category	Men's Wrist Watches
Customer Wishlist	× I just wish the hands glowed.
	× Just wish it had a metal caseback

Creativity
(100=Wildest)

0   60   100

0 10 20 30 40 50 60 70 80 90 100

↻ Generate

GENERATED IDEAS

SAVE	PRODUCT IDEA
Copy	**A carbon steel watch with a glowing face. Under each hour, you have a number of dots that light up in sequence from left to right, then back to the left again when it reaches 12.**

Figure 2.3 – Generating product ideas based on a customer wishlist

If you haven't used a language model yet, you can try out AI models such as OpenAI's GPT-3. GPT-3 was trained on billions of content pieces on the internet to generate new text, like your phone's autocomplete on steroids.

For now, these systems are pretty basic: they just generate words one after another without real intelligence or understanding what they mean. That said, GPT-3 can be used to write like a human for a variety of real-world use cases. For instance, it could be trained to produce coherent arguments in favor of a political candidate. Or it could be trained to write more evocatively about a particular memory or experience. It could also be trained to generate new product ideas.

So, while it may sound whimsical at first, there are quite real applications for this technology, which is based on an AI architecture called the *transformer*.

Transformers

GPT-3, for instance, uses the **transformer**. Transformers are the secret weapon behind some of the world's most popular natural language models, from the likes of Facebook, Google, and Microsoft.

At a technical level, transformers use what's called a **sequence-to-sequence architecture**, or **Seq2Seq**. As the name suggests, Seq2Seq is a neural network that transforms a sequence of elements, such as a string of text, into another sequence.

Seq2Seq models are made up of an *encoder* and a *decoder*. An encoder takes the sequence and turns it into a set of latent variables, which are then passed to the decoder. The decoder is responsible for turning the elements back into a sequence.

To give an example, if you have a document that says `The quick brown fox jumps over the lazy dog`, the model might like to know what animal a `dog` is referring to. This task involves encoding the word `dog` into a numerical token that can be calculated regarding other tokens.

For example, the token for the word `dog` might be located between the tokens for the words `wolf` and `pet`. Of course, this is hugely simplifying the matter, as every character and every word gets a token, creating a huge network that computers can use to understand language.

This allows computers to extract semantic features from documents in a way that had previously been thought impossible. As we mentioned previously, this model uses sentence structure as an input feature – or you could call it an *input vector* – so we can see how this works at a conceptual level here.

Sentences are parameters in supervised sequence models such as Seq2Seq; they are fed through an encoder, which turns them into vectors (latent variables). These are then passed through one or more networks before being turned back into sequences for output processing (decoding).

A quick note on terminology

These models aren't just called transformers because they transform sentences; they also transform characters into characters (or words). A character transformer doesn't just process words but also letters. It processes each character individually instead of taking groups of characters at once.

There are differences between character transformer models and word processors such as Microsoft Word or Apple Pages. However, they all do something similar under the hood: they process raw data with a computer algorithm until a result pops up on the screen that we humans can read easily.

Transformers are a crucial AI development as most people agree that robots will need excellent reading and writing skills before becoming truly useful. Whether we want robots to vacuum our floors or serve us dinner depends on how good their language comprehension capabilities are. We need robots who can understand natural language sentences in context before we teach them specific tasks.

While transformers can be trained on text, which gets encoded as numerical token values, they can also be trained on *pixels*, which get encoded as numbers (that is, a matrix of pixel RGB values). As a result, transformers can be used to make images in much the same way that they can write text, as seen in this example of a chair, which has been designed with the help of AI image generation:

Figure 2.4 – A generative design by Emmanuel Touraine, CC BY-SA 4.0 (https://creativecommons.org/licenses/by-sa/4.0)

From product designs to logo assets or other design collateral, AI-powered image generators will play an increasingly large role in the product innovation process.

Selecting product ideas

Innovation is the lifeblood of product teams. However, innovating new product ideas is just one step. Let's explore how to select the right product idea for business success.

Many business leaders understate the importance of thinking about a product as an extended process. They think that once they have a good idea, they can immediately build it and get it to market. But this line of thinking is fundamentally flawed. The most successful products in history were not developed overnight but rather evolved through systematic planning and execution processes. This basic path is how many companies sustainably grow their market share year after year.

Important questions here include the following:

- *Where should product managers focus their energy?*
- *Should they focus on the original idea (or ideas) or should they be developing products based on other opportunities that could provide an even stronger return on investment?*
- *How do you know which ideas are worth pursuing?*

Every company faces different external competitive pressures. That means every company has unique opportunities and threats associated with its products. For example, when Facebook was founded in 2004, there were no smartphones available to consumers; today, there are thousands of smartphone apps that compete with Facebook for users' attention. So, which ideas make sense for Facebook to pursue?

To help answer these questions, you can consider six key factors:

1. Your company's need.
2. What problems the product solves for your customer.
3. Whether those problems are being neglected by competitor.
4. How big the market is for solving that problem.
5. Where you are in the life cycle of a product.
6. Whether there is a new version of your product coming out soon, or whether you are currently in the market with an existing version.

The answers to these questions will help you prioritize your product ideas. For example, if your company is far along with its current product and it's not suffering from any major competitive threats, then perhaps it can afford to take some time off to work on something completely different. However, if you're still working on development and planning for a new release of your current product, then pivoting into something new may make less sense.

If you have highly skilled employees who already know how to make products like yours, they may be able to start building right away without much need for training or retraining. Alternatively, if it takes many people with complementary skill sets to build products successfully (as it does at companies such as Facebook), then you may want to expand your team first before starting a project that requires more than one person (such as hiring designers and writers). You might also consider partnering with other organizations that have relevant skills (for example, Twitter has relied on outside contractors as part of its engineering teams).

These partnerships can also help mitigate risk by spreading out development costs across multiple organizations.

This factor changes significantly, depending on where your company is in terms of fundraising activity and cash reserves. If you're raising money or looking for investors now but don't have significant revenue streams generating profits, then spending money on risky projects that aren't likely to be profitable immediately isn't so crucial at first. However, once things get going again (and assuming investors continue believing in your business model), launching products quickly will allow them to generate revenue faster.

It's important to note that having strong answers to all questions does not guarantee success; it just helps teams make better choices about which opportunities are promising enough to pursue further down the road.

Iterating product ideas

However, selecting the right product idea is not enough. Once a potentially fruitful idea has been selected, idea iteration takes over. Idea iteration is a highly strategic process.

Many companies try to execute their original product ideas without formal planning or iteration. These companies often fail to sustain momentum and ultimately kill promising but unproven products.

In contrast, many successful products have gone through explicit iterations that are guided by principles of organizational learning. While the idea itself is not changed, teams add new features, new capabilities, and new functionality in response to what users want. An example is Yahoo's acquisition of Flickr in 2005. At the time, Flickr was already a popular destination for people looking to share photos and videos with other users online. But by buying Flickr, the Yahoo team could create an even more compelling user experience by providing photo sharing.

Another example is the evolution of Facebook from the *hot or not* rating site to a global communications and media platform. Facebook started in college dorm rooms, and many early users wanted a casual way to stay in touch with their friends. But as Facebook became more popular, it attracted more professional users who wanted greater benefits from the site. In response, the team created new features such as Events and Groups that helped people find groups of friends they didn't know well but could use as sounding boards for career advice or to help them organize their workday.

Of course, not every product idea will reach such a level of success. But if your organization is going through an innovation process similar to this one, you'll want to make sure you have a strategy for iteration at each step of the way. The lesson here is that an innovation process should be guided by user wants and needs.

Summary

AI is a very broad concept. The way we use the term varies, depending on the context. In most cases, it refers to machines that can carry out difficult tasks on their own by learning from data.

Commerce.AI, as an AI platform, stands between you and your data so that you can build products that solve customers' problems more effectively than before, including by using creative AI. It captures data from every part of the product ecosystem and allows you to ask questions about products in ways a human never could, and it turns raw data into valuable information that you can use to improve your products and make better decisions about how to grow your business.

Through the pillars of language understanding, visual understanding, information extraction, and information organization, you can use AI to empower product teams with more efficient creative ideation. Creative ideation is vital for product success, but it's not the only skill you need. Industry trends change, and it takes a keen eye to see those changes coming and then pivot your product strategy accordingly. While this chapter has explored how to generate product ideas, you'll want to make sure that you're coming up with product ideas within a trending, viable market.

In the next chapter, we'll explore how to predict industry-wide trends using big data.

3
Understanding How to Predict Industry-Wide Trends Using Big Data

Forecasting is a tricky business; no one is sure why it is that some forecasts are right and others are wrong, but two main factors contribute to forecast accuracy:

- Which data and models are used
- What assumptions are made about the variables being forecasted

Unfortunately, as this chapter will show, most traditional methods of forecasting suffer from low predictive accuracy because they do not take these important factors into account properly. Here, we will explore how **big data** changes all that by enabling better predictions.

Our goal is not to present you with yet another prediction tool (though we are going to discuss a few). Instead, we want to share some insights about why conventional methods fail and how we might harness big data to make better predictions ourselves. These insights will set product teams up for success since accurate forecasts of market demand and sentiment, across product segments, are crucial for a successful launch.

To that end, this chapter will cover the following topics:

- Why traditional forecasts fail
- Using big data to enable better forecasts
- Gaining value from data-driven forecasts

Technical requirements

You can download the latest code samples for this chapter from this book's official GitHub repository at `https://github.com/PacktPublishing/AI-Powered-Commerce/tree/main/Chapter03`.

Why traditional forecasts fail

Traditional methods of generating forecasts are based on the idea that you need expert knowledge and intuition of different products and services to model their future behavior. However, this approach has fundamental limitations, as follows:

- It's impossible to know everything about all products and services.
- Knowing how products perform today is not a good guide for predicting how they will perform tomorrow.
- The behaviors of many products are highly correlated and can be difficult to disentangle.
- Traditional models get overwhelmed by today's big data.
- The data itself keeps changing.

To address these challenges, we need new forecasting methods that can handle large amounts of heterogeneous data while producing forecasts that are more reliable, more accurate, easier to interpret and explain, and more useful for decision makers.

Let's explore these limitations in detail to lay the groundwork for why new forecasting methods are needed.

It's impossible to know everything about all products and services

In a forecast, you either assume that all the relevant data was captured, or you will build in room for error. On the other hand, if you're going to try to capture everything about a product or service with machine learning models, then why not do it right from the start?

When we try to model things such as market demand for a product at a granular level using traditional statistical methods, we run into two fundamental problems:

- Statistical methods are very weak at modeling interactions between variables.

- Statistical methods are very good at capturing correlations among variables.

Conventional statistical methods simply don't cope well with the fact that many products or services interact in complex ways and share properties across different products and services (this is the basis for the popular notion of *intermediate goods*).

A prime example is airline travel: the market for flights between cities might be thought of as a series of markets within larger markets. There are flights between cities, but also flights within regions, continents, and so on. Within each market, there are many different options. This allows airlines to segment their markets into niches and then price accordingly, which explains why they often offer low fares on some routes but very high fares on others.

Most statistical models assume that all relevant variables (for example, flight duration or the number of stops) are independently distributed; they ignore how interrelated variables influence each other. As a result, they underpredict what will happen next because the statistical model assumes that customers will choose products and services based on their idiosyncratic preferences rather than on a myriad of interacting data points, including the behavior of other customers. For instance, the idea of **bandwagon consumption behavior** refers to customers with collective, conformist purchase habits across brands.

It's no wonder traditional statistical methods fail when it comes to predicting what customers will want in the future and how companies should respond to changing conditions to maintain relevance. This is particularly true today when there is more product data out there than ever before. So long as we accept these unrealistic assumptions, and build forecasts based on small datasets, our predictions will always be less accurate than reality allows us to be.

Knowing how products perform today is not a good guide for predicting how they will perform tomorrow

This is because products, like all economic variables, don't react in a simple linear fashion. Consider the traditional approach of forecasting product sales using linear regression. In this approach, you forecast product demand based on historical sales data, with minimal (or no) additional knowledge built into the model. But if your demand forecast changes because something about the market (for example, competitors entering an industry or consumers shifting their spending patterns) changes in a significant way, you end up with actual and predicted sales that are off by a large amount.

The behaviors of many products are highly correlated and can be difficult to disentangle

This is problematic for traditional forecasting because we're essentially trying to predict the future behavior of many products at once, but we still need to understand what makes any given product or feature likely to succeed or fail in the market. This is why there are so few successful product forecasts. We've seen this time and again in the media: a company releases a new product, it becomes popular for a brief period (perhaps because of some marketing campaign), and then later on we see that sales declined or that it was a niche product.

However, due to how people respond to products, the reasons behind these declines are often unclear. It could be that the product itself was no longer useful because its competitors improved their products; alternatively, it could be that consumers got bored with it and moved on to something else.

Traditional models get overwhelmed by today's big data

Traditional models weren't made to handle the modern commerce world – there's just far too much data to try to predict with simple models. In a traditional forecast, you have a bunch of different inputs (for example, sales in the past year, customer loyalty scores from surveys, and so on) and then you churn out a prediction about the future based on assumptions about these variables. The problem is: there's way too much data that goes into market success than what gets captured by these models.

This is why traditional forecasting models perform poorly at capturing big trends, whether it's the rise of e-commerce or social media or the decline of physical media: they simply can't handle today's real-time, unstructured big data from a myriad of sources.

The data itself keeps changing

As our data becomes more complex, it becomes even harder to make accurate predictions. These problems are compounded by the fact that there is no single source of data on consumer spending, nor even on consumption patterns. The US government and Federal Reserve provide some tabulations of personal income and expenditures, but those are not a complete picture. Retail stores report sales figures, but not what kinds of goods customers bought or how much they paid for them; credit card companies and banks might report high-level spending patterns, but not with granular detail.

And then there's another problem: as businesses continue to grow their online presence, it becomes harder than ever to track exactly what people buy online, as well as in physical stores. As a result, your ability to forecast is compromised by limitations in the available data that you can manually select.

Big data can help us overcome the previously outlined challenges and understand how products and services will perform over time.

Using big data to enable better forecasts

Because AI learns from big data, we can use it to uncover hidden patterns in large sets of data about separate products or entire markets and overcome the biggest challenges with traditional forecasts.

For example, with Commerce.AI, we scan billions of product data points across over 100 sources to find trending product categories, showing product teams what market opportunities are worth pursuing. The following diagram shows some trending product categories that were found using the Commerce.AI data engine. Product teams can select any category to dive deeper and find leading brands, top products, product reviews, and more.

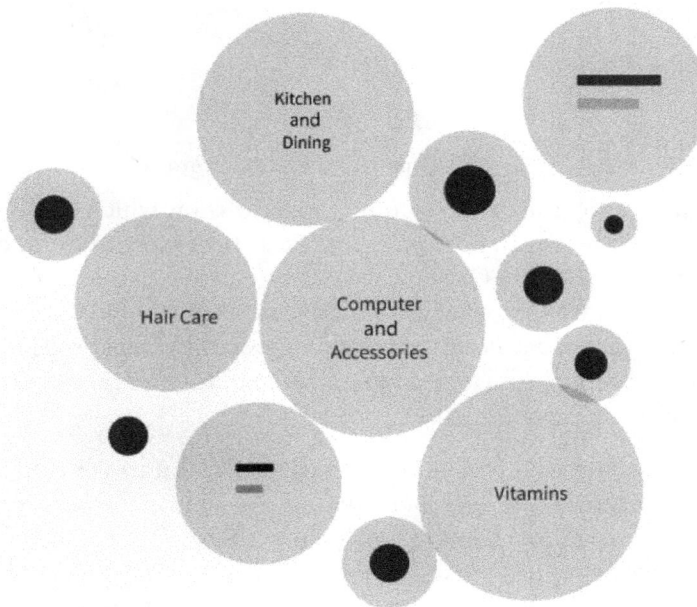

Figure 3.1 – Trending market segments

This data enables insights on specific market categories, as well as larger market segments. For instance, in the preceding diagram, we can see that markets such as computers and accessories, vitamins, and kitchen and dining are particularly large opportunities. The following diagram shows the result of an analysis that's been done on e-commerce as a whole, showing that the industry experienced 10 years' worth of growth in 3 months during the first COVID-19 pandemic lockdowns:

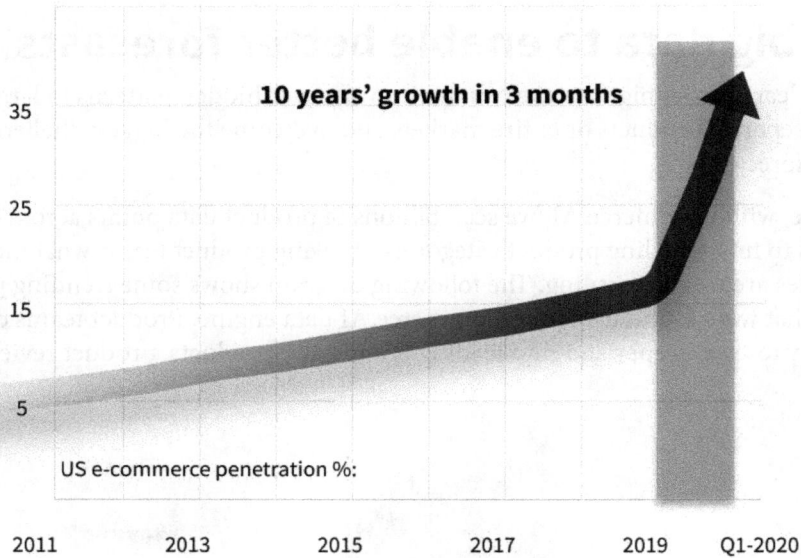

Figure 3.2 – US e-commerce penetration growth

In the case of Amazon, as an example, you might want to know which books sell best in different seasons, but also which books are popular with readers who buy other books related to food and cookery. Or you could look at historical sales volumes for different kinds of cars and wonder what factors drive consumer choices between them. With enough data, AI can yield a wealth of insights about individual products and their performance in different scenarios (for example, *What happens if I promote my new smartphone app during the Super Bowl?*).

Algorithms may even reveal correlations between particular features or distinct behaviors within a category (for example, *Which models that sold well last year share certain design characteristics?*). These insights allow us to make more informed decisions as business managers: we now have better tools than ever before for predicting the success of one product among others based on specific attributes or similar features (for example, *Do people like our latest smartphone app compared to other apps such as Uber or Lyft?*).

In particular, big data fuels an emerging, rapidly growing category of AI, known as **deep learning**, which has big implications for product teams looking to forecast industry trends.

Understanding deep learning

Deep learning is a subset of machine learning, which is the field that describes how computers can learn without being explicitly programmed. Machine learning has been around since the 1960s and has been used in many different applications, from playing video games to recognizing photos.

However, it wasn't until the 2010s that deep learning took off. That's when businesses began deploying machine learning techniques for highly complex problems such as object recognition, language translation, and computer vision using big data. Deep learning is a good fit for these more complex tasks because large neural networks have layers of artificial neurons with multiple connections between them, which helps them identify patterns in virtually anything.

Each layer *learns* some feature of the dataset, image, or part of speech by looking at examples and then adjusting the weights of its connections so that they more effectively detect similar features in future examples. The result is an extremely powerful pattern recognition algorithm: if you show it lots of examples from one category (say, cat photos), it learns to recognize cats; if you show it lots of examples from another category (say, dog photos), it learns to recognize dogs; if you show it a bunch of sales data, it'll learn to predict that too. This allows the network to generalize far beyond what any human programmer could manage on their own. So, why didn't product teams make use of this kind of technology before? It turns out that there were a couple of big obstacles researchers had to overcome to make deep neural networks work as intended.

The success of deep neural networks is due to a couple of key factors: **increasing computational power** and **vast amounts of data**.

One major challenge when it comes to making progress in machine learning is finding enough relevant data samples so that we can train our models effectively. If we look at supervised machine learning problems where labels are associated with training examples (for example, images containing dog or cat labels), then the quality and quantity of labeled training data become an issue when trying to solve any complex problem. It's common to see situations where a single dataset only covers a minuscule percentage of all possible training examples (for example, images of a certain object) in the world. It's also important to remember that training data is only one part of the machine learning pipeline – we also need to have some way of validating our machine learning models on unseen data so that we can improve their performance.

At the same time, getting this huge amount of data is vital, as AI accuracy generally increases with more data. The following diagram shows that the accuracy of natural language models increases as the number of model parameters, or, more simply, the model's size, increases. The same phenomenon is true with other types of neural networks, including those related to forecasting.

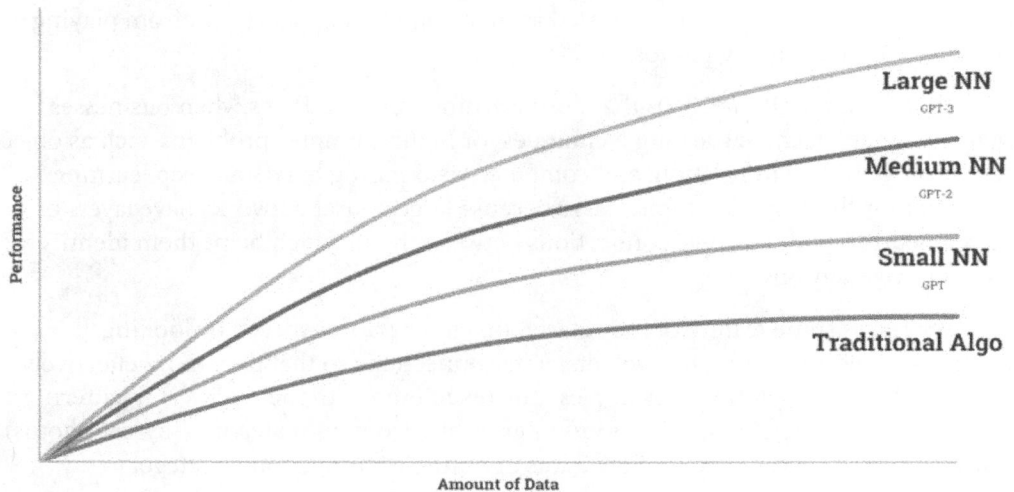

Figure 3.3 – ImageNet accuracy versus model size

However, before deep learning could be applied to huge datasets, it faced another major challenge: **computational power**. When we think about how computers work, we usually think about a bunch of logic gates (ANDs, ORs, and NOTs) that perform computations. But that's not how neural networks work, which attempt to model our brain. Our brain performs millions of calculations every second on billions of neurons – all while we're walking around or talking to someone!

So, how do we accomplish this via a computer? We need lots of computing power. And not just any kind – we need a lot of specialized hardware that can perform many different calculations at once, such as **Graphical Processing Unit (GPU)** clusters, or more recently, **Tensor Processing Units (TPUs)**. This is because machine learning models have very computationally intensive training processes that use techniques such as backpropagation. All this is needed because AI, much like us, learns from examples.

Learning from examples

Given large amounts of training data, AI systems improve their ability to make accurate predictions over time as they learn more from experience. And due mostly to recent improvements in deep learning technology, AI has gotten quite good at predicting trends based on large amounts of examples. Because it can take massive amounts of training data and build increasingly complex models using hundreds or thousands of variables, an AI system can produce very detailed forecasts that look quite realistic compared to traditional forecasts built by humans using heaps upon heaps of historical market data.

As such, machine learning platforms have been gaining popularity, and they are now widely in use among product teams, from start-ups to the Fortune 500.

Demand forecasting with a practical example

Product market fit (PMF) is an important milestone for product teams – it means that your product or service has achieved sufficient customer adoption and engagement so that it can sustain itself without additional funding from investors. PMF is usually reached when customers are sufficiently engaged with your product, have become repeat customers, and continue to purchase more of your product as their needs change over time.

To achieve PMF, you must be able to forecast demand for new products based on changing customer needs and preferences, so that you can decide whether or not it makes sense for you to invest in developing a new version of your existing product or launching a completely new one.

Demand forecasting involves determining how much demand there will be for any type of feature at any given point in time (for example, during the next quarter).

Let's walk through a practical example of forecasting demand for Adidas Yeezy sneakers:

1. First, we'll import the libraries that we need, which are Python's pandas for data manipulation and Facebook's Prophet, an AI forecasting library:

   ```
   import pandas as pd
   from fbprophet import Prophet
   ```

2. Next, we'll import our data – 5 years of worldwide search history for the term Yeezy, retrieved using Google Trends (https://trends.google.com/):

   ```
   df = pd.read_csv("multiTimeline.csv")
   ```

Here's what the associated data looks like:

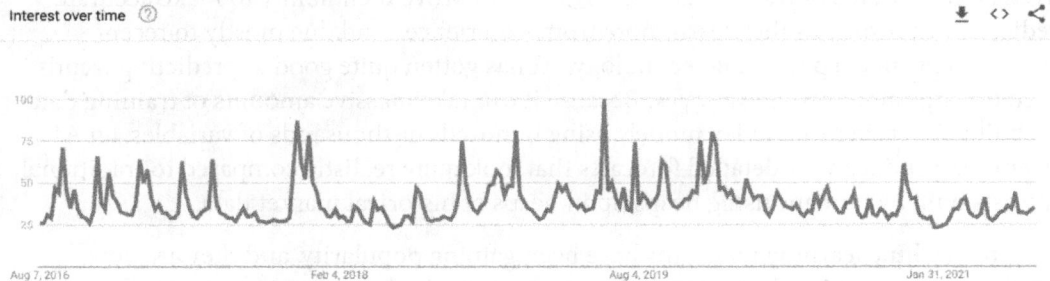

Figure 3.4 – Google Trends interest over time for the search term Yeezy

3. The preceding graph shows us search trends for the term `Yeezy`, which we now have to turn into a format that we can forecast from. Prophet requires that the datetime column is named `ds` and that the observation column is named `y`, so we'll rename both columns:

```
df = df.rename(columns = {"Week": "ds", "yeezy:
   (Worldwide)": "y"})
```

4. Building an out-of-the-box model is just two lines of code, where we first instantiate the model and then fit it to the data:

```
m = Prophet()
m.fit(df)
```

5. Now that we've built a forecasting model, it's time to make a forecast. We'll make an empty DataFrame to store forecasted values, and then fill that DataFrame with the predicted values:

```
future = m.make_future_dataframe(periods=52, freq='W')
forecast = m.predict(future)
```

6. Now, we can plot our forecast with a single line of code:

```
fig1 = m.plot(forecast)
```

With that, we've successfully predicted demand for a given product. Here's what the resulting forecast graph looks like:

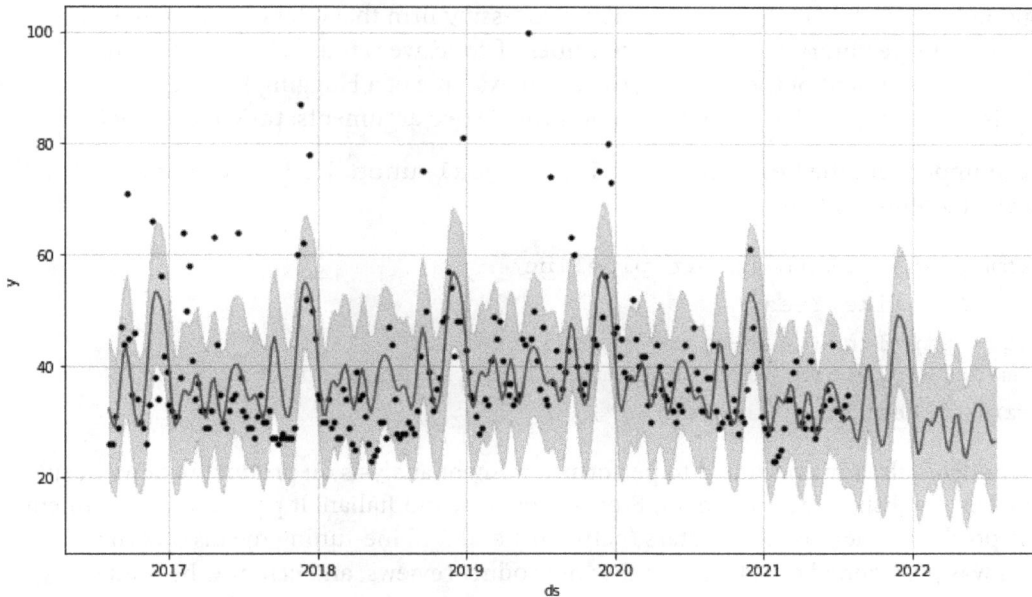

Figure 3.5 – A Facebook Prophet 1-year forecast of search demand for Yeezy

The preceding graph shows a 1-year forecast of Google Search interest for the product Yeezy. The same steps could be replicated for any product and any timespan or scaled to forecast demand across entire product lines or even market segments.

A similar method could be used to predict the sentiment of products and features over time. Combined with demand forecasts, product teams would then know the *sweet spot* of when new products and features should be released.

Sentiment forecasting with a practical example

Let's follow similar steps for forecast sentiment. After all, a product launch can only succeed if there's positive customer sentiment, and the sentiment is constantly changing, so it's crucial to be able to accurately forecast customer sentiment. If there's both positive customer sentiment and high forecasted demand, there's a potentially lucrative market opportunity at hand.

While Commerce.AI's sentiment forecasts are based on billions of data points and large, deep neural networks, we can take a simple example based on just a few data points.

Hugging Face is a leading natural language processing firm that offers a wide range of natural language libraries. With just a few lines of code, we can use Hugging Face to analyze the sentiment of the text. To provide an example of a Hugging Face sentiment analysis model in practice, we only need to provide two arguments: **task** and **model**.

We can implement the `bert-base-multilingual-uncashed-sentiment` model concerning sentiment analysis like so:

```
from transformers import pipeline
st = f"I like Yeezy"
seq = pipeline(task="text-classification",
  model='nlptown/bert-base-multilingual-uncased-sentiment')
print(f"Result: { seq(st) }")
```

This model is already fine-tuned to perform sentiment analysis on product reviews in six languages: English, Dutch, German, French, Spanish, and Italian. It predicts the sentiment of the product review as several stars (between 1 and 5). Fine-tuning means that the model was pre-trained on a large number of product reviews, and can now be used on new product data more accurately than a general-purpose language model.

AI can be used to improve our understanding of customer feedback by automatically identifying how customers feel about products they've purchased online. This allows us to understand what drives customer satisfaction within business lines – it could help us identify negative features before they become big issues for our customers, or identify areas of opportunity for new features. Another simple sentiment classification library is provided by TextBlob, which lets us run sentiment analysis in just four lines of code:

```
from textblob import TextBlob
text = "I just bought Yeezys and am absolutely in love!"
blob = TextBlob(text)
print(blob.sentiment)
```

TextBlob is an open source Python library for processing textual data. It lets you perform different operations on textual data such as noun phrase extraction, sentiment analysis, classification, and translation. By calculating sentiment on many data points, such as product reviews with a timestamp, we can graph sentiment over time.

The following diagram shows an example of the sentiment of Yeezy sneaker reviews over 1 year (the upper line is the frequency of neutral reviews, the middle line is the frequency of positive reviews, and the bottom line is the frequency of negative reviews):

Sentiment Over Time

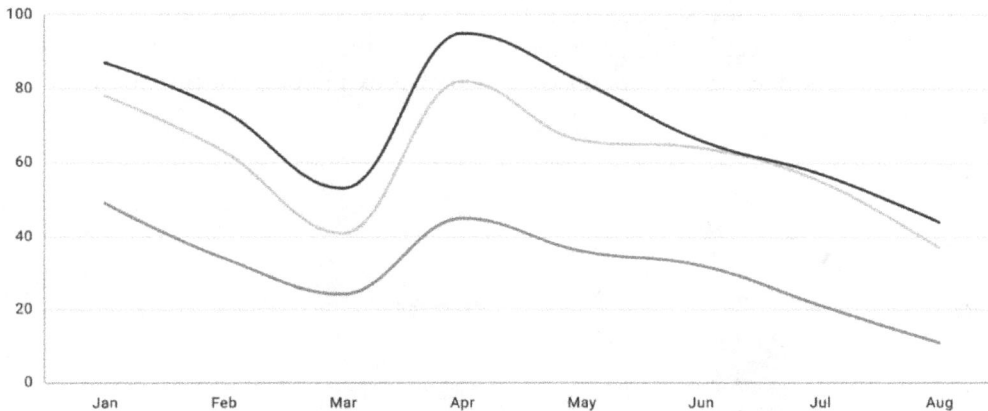

Figure 3.6 – Sentiment of Yeezy sneaker reviews over time

Now that we've looked at how to predict demand and sentiment, let's explore how businesses can gain real value from these forecasts.

Gaining value from data-driven forecasts

Product teams can gain value from any kind of forecast, whether it's a simple Google search that gives you an idea of how popular a certain product is, or something more complex, such as Commerce.AI, which allows you to look at trends based on billions of data points. Let's look at how AI can be used to provide value for product development pipelines, forecasts, and roadmaps. *Pipeline* refers to a list of all the products currently being worked on by a company or team in its **product management system** (**PMS**), such as Jira Software or Trello.

The problem with pipelines is that they can become useless if they do not change frequently enough to reflect market shifts and changing customer needs. Let's say you have 20 products in your pipeline; however, because the market might shift 2 months into the future, those products may not get shipped by then due to lack of demand. With forecasts from Commerce.AI, you can make sure that any changes are reflected immediately in your pipeline and you don't miss out on opportunities that were previously available. You can also address competitive threats as they arise, based on larger market trends. Our analysis of consumer data shows, for instance, that three-quarters of consumers tried new brands during COVID. This indicates an unprecedented threat to the market share of established brands. The following diagram shows consumer statistics extracted from the Commerce. AI data engine:

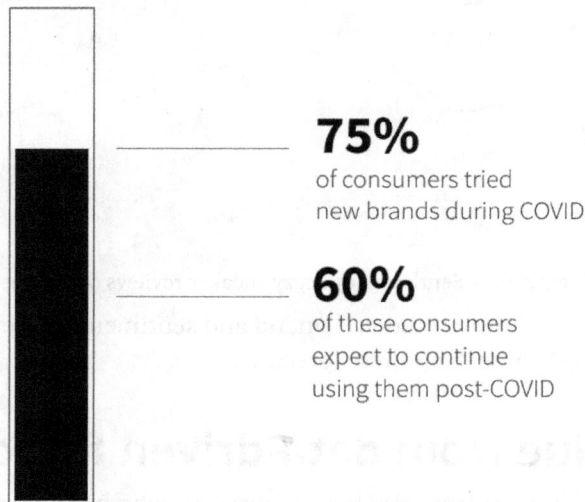

75%
of consumers tried
new brands during COVID

60%
of these consumers
expect to continue
using them post-COVID

Figure 3.7 – Consumer statistics

Using Commerce.AI also provides insight into how well launches will perform, enabling teams to create product roadmaps that are better timed, based on historical data from tens of thousands of product categories.

In short, whether it's managing pipelines or planning forecasts, having up-to-date information about current trends is essential to making high-quality decisions. Unlike traditional methods, where product teams would wait until sufficient data became available after months, such as from offline surveys using paper questionnaires, businesses can now tap into real-time insights. With today's technology, we're also no longer limited by the sources of data. AI can access many product data sources, from Amazon to Walmart to Target to YouTube video reviews.

Moreover, given that AI can scale to analyze any amount of data, product teams can find insights from data sources around the world, and they're no longer geographically limited. With these insights, they can find the perfect geographical location to launch in, and in our globalized world, that may very well be far from home.

Figure 3.8 – Map of major commerce data sources

Summary

In short, the issue with traditional forecasting is that any attempts to forecast a complex system, such as a market or economy, generally fail since there are too many variables and interactions among them for any simple model to capture them all. Not only does AI deliver better predictions within a product segment, but it's also more effective across industries, as it can scale to learn from any amount of data, anywhere in the world. With knowledge of how to forecast industry-wide trends, product professionals can ensure that they don't waste time on market opportunities that won't pan out, and instead focus on the red-hot opportunities that will drive business success.

In the next chapter, we'll dig into practical examples in the industry, and look specifically at how top luxury goods brands are using data and AI to power product success.

Section 2:
How Top Brands Use Artificial Intelligence

In this section, you will gain a deep understanding of how top commerce and product brands are using AI tools such as Commerce.AI to drive innovation.

This section comprises the following chapters:

- *Chapter 4, Applying AI for Innovation – Luxury Goods Deep Dive*
- *Chapter 5, Applying AI for Innovation – Wireless Networking Deep Dive*
- *Chapter 6, Applying AI for Innovation – Consumer Electronics Deep Dive*
- *Chapter 7, Applying AI for Innovation – Restaurants Deep Dive*
- *Chapter 8, Applying AI for Innovation – Consumer Goods Deep Dive*

4

Applying AI for Innovation – Luxury Goods Deep Dive

Luxury goods are associated with exclusivity and prestige, which makes product development particularly tough. There's a far smaller, more exclusive market of people to target, and traditional methods of market research, such as focus groups, fall short in luxury markets, which are largely made up of **high-net-worth individuals** (**HNWIs**).

Moreover, the value of a luxury goods company comes largely from its brand, so analyzing brand data is of special importance. At the same time, luxury brands naturally have far fewer customers than mass-market brands, which means that getting that data through normal means proves to be a real challenge.

In this chapter, you'll learn about the following topics:

- Understanding the challenges of luxury brands
- Understanding the data extraction process
- Using Commerce.AI for luxury brands

Whether or not you're operating in the luxury market, this chapter will give you insights into how specific brands operate, the challenges they face, and how they overcome them with data and AI, which has broad importance for any brand seeking a competitive advantage.

Technical requirements

You can download the latest code samples for this chapter from this book's official GitHub repository at `https://github.com/PacktPublishing/AI-Powered-Commerce/tree/main/Chapter04`.

Understanding the challenges of luxury brands

Luxury brands stand out from other competitors because of their high prices: they command higher margins than mass-market rivals due to higher brand equity. They also offer distinctive aesthetic experiences that go beyond mere functionality: beautiful clothing or jewelry that is designed to convey a desirable image for consumers (and hence, raise brand equity).

These attractive products appeal to customers' aspirations for style and status. The typical consumer perceives luxury goods as just above their budget range but still within their reach and, therefore, desirable as a status symbol, as long as the brand successfully maintains its image. As a result, brand management is of unparalleled importance for luxury brands.

Brand management

While damage to one's business reputation is harmful in any industry, this is particularly the case for luxury brands, where brand image is the make-or-break quality of the business.

For example, the vast majority (78%) of Chinese consumers will buy or boycott a brand solely because of its position on a societal issue, as revealed by the *2018 Edelman Earned Brand* survey.

Brands that fail to collect, analyze, and act on consumer data will inevitably run into anti-brand sentiment.

Increasing competition

Another key challenge is that the luxury market is highly competitive, with many players fighting for the same customers. This makes it very difficult for luxury brands to differentiate themselves from one another.

In order to stand out from their competitors and maintain their brand equity, luxury brands need to attract and retain customers through a comprehensive understanding of the customer experience. This is especially true in the context of online channels, where customers have the ability to compare products and services across multiple channels and sources.

The term *luxury* has become increasingly popular over the past decade, as consumers have become more willing to pay for products with higher prices. As wealth increases around the world, demand for luxury goods has been growing steadily too.

While most Americans still prefer cheaper products (such as grocery store brands), wealthy consumers are showing an increasing preference toward premium products such as Birkin bags from Hermes or key chain necklaces from Dolce & Gabbana.

The growth of luxury markets can be traced back through centuries of economic development. During the Roman Empire, when people were wealthier than ever before, they spent more on food and wine than ever; during this time period, winemaking became highly developed as a result of increased demand for fine wines.

At the end of World War II, there was a similar increase in demand for fine suits and other clothing due to mass production techniques used by American factories during wartime needs being applied to civilian industries such as clothing manufacturing.

Today's rapid growth in wealth can be traced back even further than that: many historians believe that it was during prehistoric times when humans first started trading valuable items with each other based on perceived value rather than actual value (such as shells or stones).

In short, the luxury market has a very old history, but it has only more recently truly taken off, resulting in stiff competition.

Growth is driving competition

The luxury market is a $300 billion industry that includes everything from handbags to yachts, and it's growing at a rapid pace (`https://www.statista.com/study/61582/in-depth-luxury/`). And this growth isn't expected to slow down anytime soon: according to Statista, global luxury goods sales are expected to grow by 5.4 percent per year, reaching nearly $400 billion by 2025.

This growth has been fueled by several factors, including rising wealth around the world and mass manufacturing techniques that have made it easier for companies to produce products with a high profit margin.

But perhaps the biggest driver of this growth is social media. As consumers increasingly turn to social media sites such as Facebook and Instagram for fashion inspiration, brands have been able to use these platforms as marketing tools. They can post photos of their latest creations on these sites and get them seen by millions of people who could potentially become customers.

Social media has also helped brands reach out to younger generations who are less likely than older generations to frequent traditional retail stores or visit physical boutiques. Instead, they prefer shopping online or via mobile apps, where they can see more product options at once and compare prices easily without having to physically visit a store first.

Social media management

Because of this shift toward online shopping, many luxury retailers have had trouble keeping up with demand for their products as shoppers increasingly seek out cheaper alternatives online instead of paying top dollar for designer goods in physical stores or boutiques.

This means there's plenty of room for new players in the luxury market who don't necessarily need deep pockets but do have an eye for design and know how to use social media effectively as marketing tools.

The luxury market is a tough one for any brand to crack. It's not just that the products are expensive, but that they also have a reputation for being ostentatious and frivolous. That means luxury brands must walk a fine line between appearing exclusive and appearing tacky. And it's not easy, with many luxury socks, for instance, costing thousands of dollars.

But there are some ways in which luxury brands can be more successful than their mass counterparts on social media. For example, they tend to have more sophisticated strategies for managing their online reputations and building relationships with customers. They also have access to data about their customers that other companies don't, allowing them to personalize marketing campaigns and engage with customers in new ways.

Let's take a closer look at how these strategies work and what makes them so effective. Luxury brands tend to do two things on social media: they post photos of beautiful products or events, and they share stories about the people who wear those products or attend those events.

This strategy is similar to how many mass retailers use Instagram: post photos of your latest sale items, and then tag them with relevant hashtags so your followers can find them easily when browsing through their feed.

But this approach has its limits: luxury brands can only reach so many qualified leads through generic social media posts; if you want your followers to find your posts organically (that is, without following you specifically), you need to make sure those posts get seen by as many people as possible.

Matching eccentric customer preferences

Luxury brands face many challenges in delivering what customers want. While mass-market brands focus on addressing tangible things such as feature requests, luxury brands must also deal with the intangible.

The luxury market is very different from the mass market. It's not about what you can see or touch; it's about how you feel and what you think. It's about emotions and feelings. This means that luxury brands must be more creative in their thinking than mass-market brands when it comes to delivering value to customers.

And this is where things get tricky: *how do you satisfy an emotional need?* In other words, *how do you make people feel good?* The answer for many luxury brands is that they have to become experts at understanding their customers and then satisfying their needs through storytelling.

There are some challenges in doing so: one challenge is that we know less about our consumers than we used to because they are becoming more diverse and internationalized.

We don't know who they are or where they come from; *we don't even know if they exist! So how do we reach them? How do we find them? How do we engage with them? What kind of information should we collect from them? How can we use this information to create better products and services for them?*

This means that luxury brands have to rely on customer research more than ever before, which can be difficult when customers aren't willing to share much information.

As a result, these companies often turn to celebrities as brand ambassadors instead of relying on traditional surveys and focus groups. But while celebrity endorsements may seem like a simple solution for luxury companies looking for ways to connect with customers, there are some risks involved here too: for one thing, celebrities tend not to be representative of the general population, so using them as spokespeople could alienate potential customers who don't look like Paris Hilton or Kim Kardashian.

Celebrity endorsements also carry a lot of baggage around image issues, which could hurt the brand overall if the celebrity endorser ends up tarnishing its reputation over time.

So, what does all this mean for luxury consumers? It means that luxury consumers will continue spending money on experiences rather than tangible goods. They want something unique and memorable rather than something ordinary and cheap. They want something exclusive rather than something common and available everywhere else, and they want something personal.

They want stories instead of statistics; anecdotes instead of datasheets; inspiration instead of practical advice; romance instead of sex appeal; mystery instead of transparency; magic instead of logic; dreams instead of reality checks; surprises instead of predictability; adventure rather than security; and fantasy rather than practicality.

Understanding unique customer profiles

The challenges of luxury brands include understanding unique customer profiles. Let's explore four unique luxury customer profiles in detail, and the implications of these different profiles on product innovation: **patricians**, **parvenus**, **poseurs**, and **proletariat**.

These four wealthy consumer profiles were categorized in a famous *Wall Street Journal* article from 2010, titled *The Four Species of Wealthy Consumers*.

Patricians

The **patrician** customer profile is a heritage category, which means they have money to burn. They want luxury brands to be different and stand out in their own way. In other words, to target patricians, brands need to create *distinctions worth bragging about*.

At the same time, they don't want things that are too flashy or aggressive because those may attract unwanted attention. What they do want is something that is distinctive and exclusive without being ostentatious or brash.

In other words, patricians are more interested in the iconic than the ephemeral. While trendsetting designs are often ostentatious, iconic designs are largely more subtle, enabling them to stand the test of time and changing consumer tastes.

This understanding is crucial for product innovation because if your target market consists largely of patricians, creating ephemeral designs will result in failure. Instead, you'll want to learn from iconic products and innovate luxury concepts that push the envelope while maintaining class and delightful *understatedness*.

Parvenus

Parvenus are seen as relatively uncouth status seekers who spend lavishly on status goods such as glittery apparel, shiny cars, and gold watches. They are willing to pay a premium for the best because they are focused on their own status symbol: *what I wear defines me.*

In fact, this is true of all luxury consumers; however, parvenus have a unique focus on what others think of them. Therefore, they will often look to emulate celebrities (or people whom they perceive as celebrities) rather than achieving status through accomplishments.

As a result, parvenus tend to emphasize youth in their physical appearance; they want to be young forever. They also desire products that convey youthfulness and energy. The result is that fashion brands targeting this class frequently use celebrity endorsements and catwalk shows to project youth and energy into parvenu shoppers' consciousnesses.

The upshot is that parvenus lack patience; they expect immediate gratification from luxury purchases even if the benefits do not materialize immediately. That said, once these consumers make a purchase, they do tend to be loyal customers because once given a taste of luxury, many parvenus cannot resist returning for more.

Poseurs

How do you market to poseurs? A **poseur** is a person who wants to wear luxury goods, but often can't afford the real thing, so they spend outside their budget, acquire knock-offs, or buy the #3 or #4 brand instead of the market leader.

They want designer labels at bargain prices. And they have short memories. It is a little like marketing for a diet product: people know that diets largely don't work, so when you tell them they can eat what they love again, it doesn't matter if it has artificial sugars or preservatives or whatever else in it; all that matters is that it tastes good.

But there's also another side to this phenomenon: brands create items that let these people feel like they belong among the privileged elite without actually buying into their lifestyle and values.

This is how there are luxury goods so universally beloved by poseurs, because they signal ostentatious wealth even though their quality is generally poor compared with authentic luxury items.

To most luxury consumers, the use of knock-offs would be an insult; however, these kinds of fakes have gained enough interest within certain circles that some genuine fashion houses now authorize them to make replicas for exclusive sale only at certain stores around the world.

Proletariat

Finally, the proletariat are customers who are less affluent but status conscious. These customers have a strong need to demonstrate their social position and wealth in order to feel comfortable in their social circles and at work. They do so by showing off their homes, clothes, cars, watches, and so on.

In other words, those who have not reached the higher rungs of the economic ladder still continue to follow consumerist values of conspicuous consumption in order to maintain a sense of status quo with regard to their place in society.

In short, traditional marketing, product development, and brand management fall short for luxury brands, which face a unique set of challenges, given their unique target market.

Let's explore how analyzing product data can help luxury brands overcome these challenges.

Understanding the data extraction process

A lot of these companies use AI to predict what consumers want. They use data to understand how consumers actually behave and then use that information to shape the products they make, sell, and market. These companies have figured out how to use data in a way that provides massive value for their customers. The following examples demonstrate this in action.

Tumi uses AI for better marketing

Tumi is a manufacturer of high-end suitcases and bags for travel. Brands in this market segment are often stuck with boring marketing strategies centered around all-too-typical tactics such as flash sales and promo codes.

Tumi, in contrast, has used AI to personalize messages in its outbound marketing and smarten its digital ads strategy by targeting customers with the highest predicted lifetime value. The majority of Tumi's marketing revenue comes from its email campaigns, so improving emails with personalization was a clear way to boost the bottom line.

Given data such as recent browser behavior, email open rates, and search behavior, in addition to past purchases, machine learning models can accurately predict what a customer is most likely to buy, and when. By presenting those recommendations, Tumi was able to increase sales with fewer outbound messages.

Burberry uses AI to improve its clothes

Burberry is famous for its leather trench coats, cashmere scarves, and other luxurious designs. However, creating new styles for each season isn't easy, as it takes months for designers to create new collections based on consumer feedback gathered at stores around the world — which meant that stores often got stuck selling old styles all year long instead of offering new ones throughout the year.

Now, Burberry uses AI to improve its product innovation process, through making better products, faster, cheaper processes, and more insightful analysis.

This British fashion label also offers data-driven personalized product recommendations, both online and in-store, resulting in a 50 percent increase in repeat purchases by 2015. Additionally, Burberry uses chatbots powered by natural language as a smart *fashion adviser*.

While the exact workings of these fashion advisers are secret, as they are, after all, a valuable intellectual property, we can see how they might work. We can create a simple fashion adviser powered by **natural language processing** (NLP) algorithms using **GPT-J**.

GPT-J is a large language model, or a machine learning model that was trained on large amounts of text, released by a group called Eleuther AI. We'll demonstrate it as follows as it's an open source and easily accessible tool:

1. First, we can install GPT-J and then import our needed library, like so:

   ```
   !pip install gptj
   from GPTJ.Basic_api import SimpleCompletion
   ```

   Just like that, we're ready to use GPT-J. The way that modern large language models work is with a concept called **pre-training**, which means that the model was already trained on a huge amount of text data and we only need a small amount of data to tune the model for a specific task.

2. Next, we define this task by providing `prompt`, which includes examples of fashion advice, such as recommending the pairing of a blue denim jacket with golden yellow eyeshadow:

   ```
   prompt = "Recommend a fashion item based on a list of
   clothes item.\n##\nWearing: Floral skirt\nFashion item:
   Try a black-and-white polkadot shirt for the classic
   floral and dot combo!\n##\nWearing: Blue denim jacket\
   nFashion item: Try golden yellow eye-shadow for a warm
   blue and gold glow!\n##\nWearing: Red lipstick\nFashion
   item: Pair your red lipstick with a red dress and red
   heels for a triple threat!\n##\nWearing: " + item + "\
   nFashion item:"
   ```

This prompt is crucial for the language model because language models have a broad range of use cases, including classification, generation, translation, transformation, and so on. So, they have to be guided to complete specific tasks. Using the `prompt` variable, we can guide the model to act as a fashion adviser.

3. Further, we'll want to pass a number of parameters, primarily `temperature` (or randomness), `max_length` (or the maximum output size of the model), and `item` (or what the user types in, such as `Orange shorts`):

```
temperature = 0.4
top_probability = 1.0
max_length = 15
item = "Orange shorts"
```

4. Finally, we can now pass the `prompt` variable and the parameters to the model to create a recommendation. We'll also grab just the first line of text generated, in case the model goes overboard:

```
query = SimpleCompletion(prompt, length=max_length,
    t=temperature, top=top_probability)
Query = query.simple_completion()
lines = Query.splitlines()
results = []
```

In doing so, giving the model an input such as `Orange shorts` generates a recommendation such as **Try a dark blue top and orange heels for a bright pop of color!**

There are other ways to try this same concept out even without using any code at all, such as with **AI21 Studio** (`https://studio.ai21.com`). In *Figure 4.1*, we use the same prompt in a visual canvas instead of code and given the item `Orange shorts`, AI21 Studio recommends we **Try a floral top and sandals for a cute, summery look!**

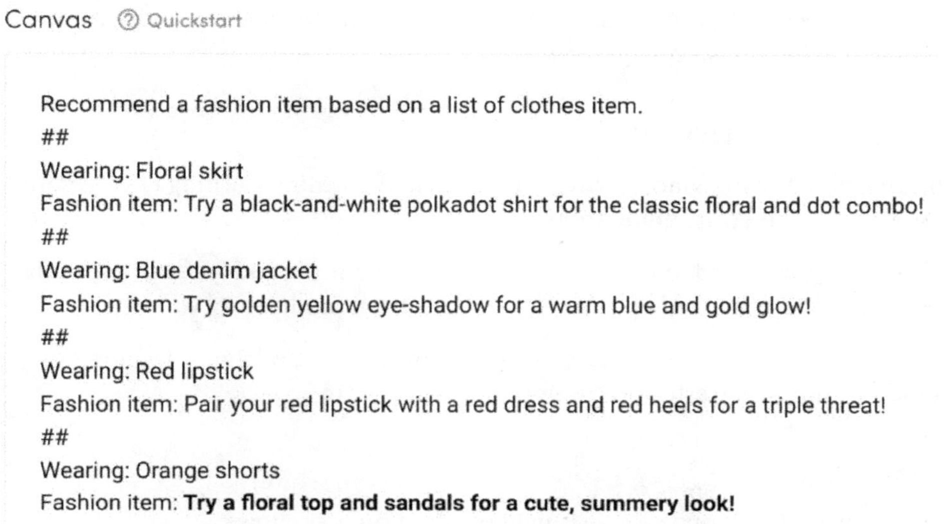

Canvas   ⑦ Quickstart

Recommend a fashion item based on a list of clothes item.
##
Wearing: Floral skirt
Fashion item: Try a black-and-white polkadot shirt for the classic floral and dot combo!
##
Wearing: Blue denim jacket
Fashion item: Try golden yellow eye-shadow for a warm blue and gold glow!
##
Wearing: Red lipstick
Fashion item: Pair your red lipstick with a red dress and red heels for a triple threat!
##
Wearing: Orange shorts
Fashion item: **Try a floral top and sandals for a cute, summery look!**

Figure 4.1 – The AI21 Studio canvas as a fashion adviser

As with GPT-J, we'll need to provide a number of settings, which is done in AI21 Studio through a **Configuration** panel.

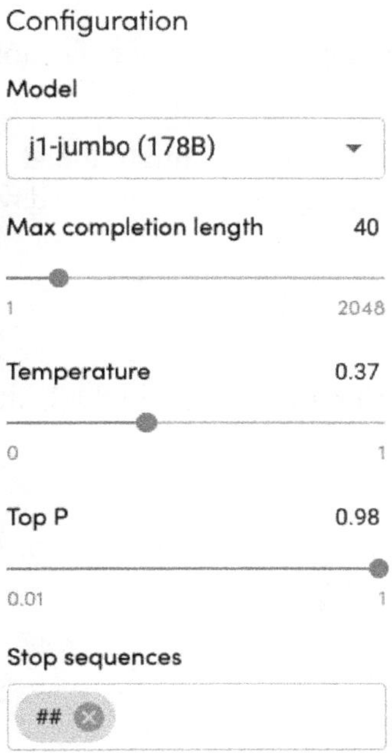

Configuration

Model

j1-jumbo (178B)   ▾

Max completion length   40

1   2048

Temperature   0.37

0   1

Top P   0.98

0.01   1

Stop sequences

✕

Figure 4.2 – The AI21 Studio Configuration panel

The settings, as seen in *Figure 4.2*, are almost identical and include maximum completion length, temperature, and stop sequences.

Algorithmic couture

Fashion designers Kotaro Sano, Kazuya Kawasaki, and machine learning engineer Yusuke Fujihira teamed up to create **algorithmic couture**.

Machine learning was used to learn from massive amounts of product data, particularly fashion images, and generated optimized fashion pattern modules.

Using computer-aided design software, the team could then model new fashion designs that would be both zero waste and comfortable.

AI runways

Another popular AI fashion use case is in generating runway shows.

AI artist Robbie Barat has worked on an AI-generated Balenciaga runway show with the help of a neural network trained on past collections of the fashion brand Acne Studios.

RefaceAI

AI is also used on the consumer and marketing side of fashion by using **deepfake technology** to place consumers in the content of brands.

The RefaceAI app, for instance, swaps the user into branded videos. They've generated over a million *refaces* and have gained 400,000 shares in a day during a test collaboration with Gucci.

Zalando

Zalando Research is an online retailer in Europe and the United Kingdom, and they're developing AI software for designers.

These AI solutions cover the personalization of fit, visual searches for fashion images, determining the diversity of design, fashion purchase recommendations, generative fashion design, and more.

Given that research advancements in AI fashion are so hot, we can expect the benefits of AI to continue to trickle into the industry for years to come. This is especially true considering the rise of easy-to-use AI for commerce tools, such as Commerce.AI, which we'll explore next.

Using Commerce.AI for luxury brands

Commerce.AI's solutions help clients gain a competitive advantage by providing them with actionable insights that enable them to make more informed decisions and improve the overall customer experience. This allows brands to connect with their customers in new ways, which in turn drives growth and profitability.

Let's dive deep into four areas where Commerce.AI is used.

Design and user research

Commerce.AI delivers insights into consumer behavior that help companies develop products that are better aligned with changing needs and preferences. Our technology helps brands understand their customers' behavior by learning more about what they buy and how they buy it.

For example, we can help luxury brands such as Gucci or Burberry understand the preferences of their most valuable customers so they can deliver personalized experiences to those consumers across all touchpoints – from product design to shipping and handling to returns.

We also help food and beverage companies learn more about the eating habits of their most loyal consumers so they can develop products that appeal to specific segments of their target audience, such as millennials who love sweet flavors but refuse artificial ingredients.

With this knowledge at hand, businesses can make smarter decisions on where to invest in innovation efforts and how best to reach out to consumers based on behavior uncovered through data analysis.

Product development and marketing

Let's say you are a luxury brand that makes handbags or luggage that you sell directly to consumers online or at your retail stores (or both). You have an existing product line that you want to expand into new markets such as China or India where there's huge potential for growth, but your current team doesn't have experience developing products for these markets.

Going blindly into a new market is a recipe for failure, particularly in the luxury brand industry, where, as we've explored, there are wholly unique customer profiles and preferences to consider. Product data enables a diverse range of insights, depending on the market niche, the brand, the customer profile, and more. As such, Commerce.AI is essentially a *blank slate* for brands – a mock-up of which is shown in *Figure 4.3* – which is filled with relevant data for the needs at hand:

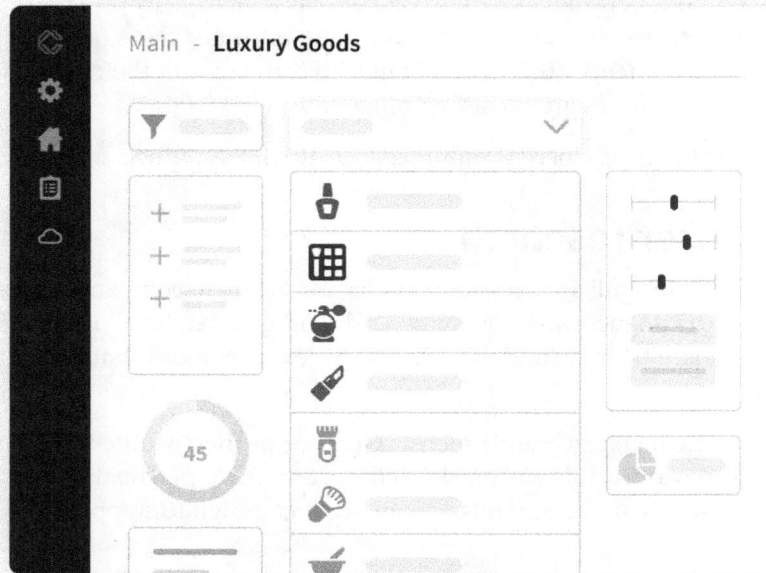

Figure 4.3 – A mock-up of a Commerce.AI Luxury Goods dashboard

The first step in this process is helping your team identify which features are most important in these new countries based on insights from industry experts about what customers really value about your products in each country.

With Commerce.AI's data engine, which has analyzed billions of product data points across geographies worldwide, you can pinpoint how customer sentiment and interests vary in different markets. For example, in *Figure 4.4*, we can see an example of popular brand attributes for a leading clothing brand:

POPULAR ATTRIBUTES

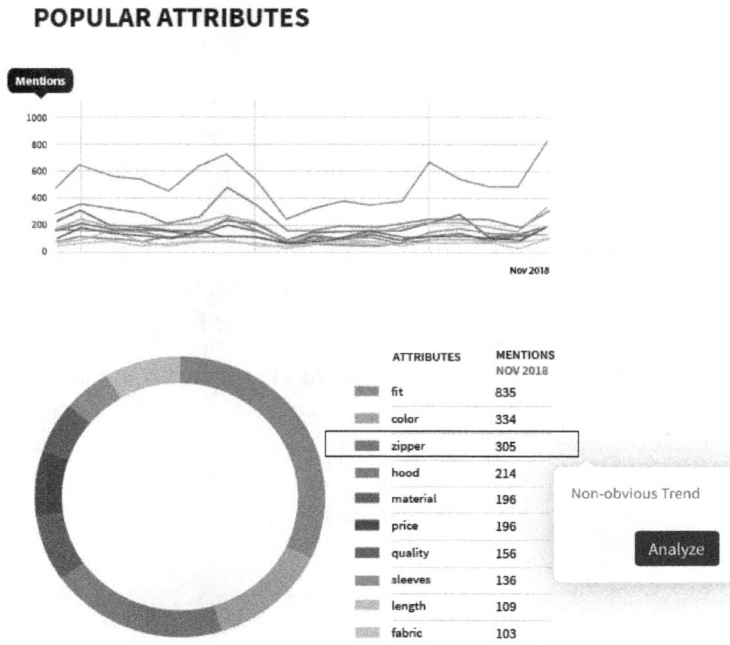

ATTRIBUTES	MENTIONS NOV 2018
fit	835
color	334
zipper	305
hood	214
material	196
price	196
quality	156
sleeves	136
length	109
fabric	103

Non-obvious Trend

Analyze

Figure 4.4 – An example of luxury brand attribute analysis

Popular attributes, or features of products, include mentions of **fit**, **color**, and **zipper**. The **zipper** attribute is highlighted as a non-obvious trend, because, for instance, customers may have issues with the zipper in particular products. By highlighting popular attributes, AI can help product teams focus on what customers are talking about the most.

Similarly, once your team has identified the most important product features for each country, it will then be able to create a prioritized list of feature ideas that they can test with actual customers across different geographies, while also validating whether customers actually want these additional features through an online survey tool built into Commerce.AI, called **voice surveys**.

Once validated by real users across different geographies using voice surveys, they can then decide which ones make sense to pursue further by creating mock-ups of how those new features might look.

Brand management

Commerce.AI helps luxury firms manage their brand in the market.

It uses machine learning to help companies understand how customers are interacting with their products, what they like about them, and what they don't like about them. The technology can be used to monitor competitors' products and identify new opportunities for growth.

For example, if you sell watches, Commerce.AI can tell you which watch styles are selling well and which ones aren't. You can then use this information to create more watches of those types or develop new watch designs based on customer feedback.

Figure 4.5 shows a mock-up of what these forecasts might look like. While brand-specific data is largely private and therefore cannot be shared here, Commerce.AI users see tremendous value in forecasting brand trends

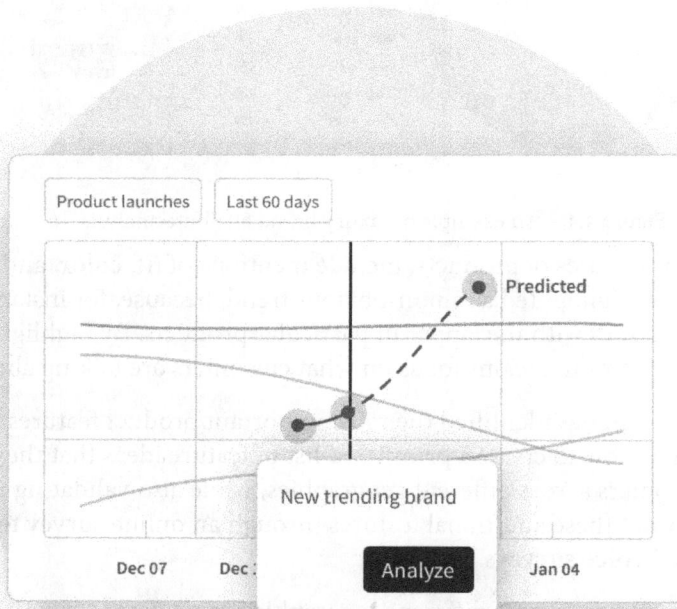

Figure 4.5 – A mock-up forecast of luxury brand trends

Commerce.AI also allows luxury brands to track features that customers want in their products and identify areas where they need improvement. This allows them to respond quickly when customers have issues with their products or request new features through social media or other channels.

For example, if a customer posts a negative review on Facebook about your watch brand quality, Commerce.AI can notify your brand team so that they can address the issue immediately before it becomes a bigger problem for your brand reputation.

In addition to helping luxury brands improve their product development process, Commerce.AI also helps them scale communications efforts across all of their channels (social media, email marketing lists, and others) so that they can respond quickly to any customer complaints or questions about their products.

Trend analysis

Luxury brands are tasked with leading the way when it comes to status trends, and that means they need to be deeply in tune with the pulse of the luxury market.

With Commerce.AI's data engine, we've uncovered a number of key trends that have big implications for product innovation, including the rise of second-hand luxury goods, the rise of luxury rentals, and luxury subscriptions. Let's explore these three trends in detail.

Second-hand luxury goods

Commerce.AI data shows that second-hand sales of luxury goods saw an astonishing rise in the past year, even while luxury sales took a hit during the COVID-19 pandemic.

As IMARC Group reports, second-hand luxury sales are expected to grow annually at a rate of over 10% in the next 5 years (`https://www.imarcgroup.com/secondhand-luxury-goods-market`). In other words, the second-hand luxury goods market is growing far faster than its core market counterpart.

This represents a massive shift in consumer preferences, which is why many brands are innovating to meet these new demands.

Luxury rentals

The luxury rental space is gaining momentum as luxury consumers become more eco-conscious, and fewer consumers want to purchase a good that they'll only sport once or twice.

This has created a new demand for rental services across jewelry sets and art collections that can be rented by the night or day for events such as weddings or corporate meetings.

Luxury subscriptions

Online retail sites have elevated digital subscription models beyond books and music, streaming into other categories such as fashion where you can subscribe to monthly deliveries straight to your door.

With services such as Birchbox, CURATEUR, and others continuing to gain traction in this space, we expect consumers will become increasingly comfortable with periodic fashion deliveries right into their homes.

Summary

In this chapter, we've explored how AI has the potential to help luxury brands overcome key product-related challenges, including brand and social media management, meeting eccentric customer preferences, and standing out amid an increasingly competitive market.

In a traditional product intelligence environment, analysts must manually sort through copious amounts of data to find market opportunities, analyze competitors, and create innovative new products.

With an AI solution, billions of products data points can be analyzed effortlessly to create a competitive advantage in the luxury market. By understanding challenges and AI-based solutions in the luxury market, you can begin to see how AI can be useful for any product team.

In the next chapter, we'll explore how to apply AI for innovation in the wireless networking space. As with luxury brands, wireless networking firms face a unique set of product-related challenges, which can be addressed with AI due to its versatile ability to learn from large amounts of product data.

5
Applying AI for Innovation – Wireless Networking Deep Dive

Wireless networking brands are facing increasing competition and ever more demanding users, all while handling unprecedented traffic. Users are demanding 5G, sustainable devices, and of course, constant connectivity. To meet these needs and gain competitive advantages, wireless networking product teams need to analyze data at scale.

Commerce.AI finds opportunities for product innovation by mapping customer needs with each brand's offering. This chapter will outline the challenges wireless networking brands face, which Commerce.AI is built to solve by identifying the key benefits of our product data engine.

Further, we will show how wireless networking brands can leverage Commerce. AI's product data engine to identify new opportunities for innovation that help them understand what's driving the needs of their customers.

In this chapter, you'll learn about the following topics:

- Understanding the challenges of wireless networking brands
- Analyzing the product data for wireless networking brands
- Using Commerce.AI for wireless networking brands

Whether or not you're operating in the wireless networking market, this chapter will give you insights into how specific brands operate, the challenges they face, and how they overcome them with data and AI, which has broad importance for any brand seeking a competitive advantage.

Technical requirements

You can download the latest code samples for this chapter from this book's official GitHub repository at `https://github.com/PacktPublishing/AI-Powered-Commerce/tree/main/Chapter05`.

Understanding the challenges of wireless networking brands

Let's start by exploring economic and technological challenges in the wireless networking industry, and how they affect the way wireless networking brands are positioned in the marketplace. This will lay the framework for understanding why data-driven and AI-powered solutions are needed.

In particular, we'll discuss challenges related to the growth in wireless networking traffic and associated performance challenges, as well as increasing ecosystem complexity, and needs for sustainability, 5G, and becoming data-driven.

Growth in traffic

The wireless networking industry is undergoing a period of significant change. The growth in mobile usage and **Internet of Things (IoT)** devices, as well as the demand for hotspots, has created an unprecedented surge in demand for connectivity – and the need for new ways to meet this demand.

The number of unique users connecting to public Wi-Fi hotspots has quadrupled over the past 5 years, according to Statista (`https://www.statista.com/statistics/677108/global-public-wi-fi-hotspots/`). Since 2019, there have been more mobile devices than people.

However, this growth has come at a cost to the operators of these hotspots, who must invest heavily in infrastructure to handle the increased demand. Wireless networking operators are increasingly responsible for providing a reliable and scalable network that will meet the growing demand of users.

Performance challenges

This growth in traffic has important implications for network performance. Network performance continues to be an issue that plagues both wired and wireless networks globally because there are only so many antennas available for signal amplification, satellites are expensive to launch into space, and cell towers must contend with interference from other wireless networks.

However, these limitations don't mean that network performance has been stagnant; quite the contrary. We continue to see advancements in network architecture thanks in large part to advancements in **software-defined networking (SDN)**, which creates virtual networks within an existing physical infrastructure using software, instead of hardware components such as switches or routers.

SDN allows networks operators to dynamically shape traffic patterns across connected devices without impacting end-user experiences such as speed or latency. This flexibility allows operators greater control over how they utilize their network infrastructure which can then be used strategically – think about allocating resources to certain types of applications rather than having applications compete against each other for limited bandwidth resources when running simultaneously on one device, network connection, or geographic region.

With this level of control comes great responsibility; providers must make sure that users aren't negatively impacted during times when demand outweighs available capacity, which results in slower download speeds or dropped connections.

This unprecedented rise in demand for wireless connectivity also poses a huge opportunity for brands that are able to capitalize on it with innovative products and services tailored around consumers' specific needs and preferences.

What does this mean for you as a brand? It means that if you can figure out how to best serve your customers' wireless connectivity needs, they could become lifelong customers who buy into your brand and its values every time they connect wirelessly – especially if you can provide them with something better than what they'd get from competing brands or non-branded alternatives. This valuable information can be extracted from product and consumer data, including product reviews from sources such as Amazon, Walmart, and Target.

And since today's customers are tomorrow's referred customers, investing in data and AI now could pay off big later, when these future customers become loyal brand advocates and evangelists who can help bring in more new business down the line.

Increasing complexity

Another major challenge brands are facing is the increasing complexity of the wireless connectivity ecosystem.

When you think about it, there are more parts in an electronic device (such as transistors in a chip) than there are in an automobile engine. With each passing year, these pieces become increasingly important as they influence how well our devices function.

At the same time, consumers have more options than ever when it comes to choosing which devices they'd like to use for specific applications or functions. For instance, tablets and laptops can be used for productivity purposes, while smartphones are more often used for social media or entertainment. Meanwhile, wearables can be used for health monitoring, smart speakers can be used for listening to podcasts or streaming music, and VR headsets can be used for immersive experiences. The list goes on and on.

So how do you figure out what your customers need – and want – when it comes to connecting wirelessly? It will require a significant investment of both time and resources as you identify what your target customer wants, needs, values, and dreams of for their own personal wireless experience.

Then, you'll need to figure out how best to provide them with that experience through new products and services that meet their needs within the context of your brand's overall value proposition.

Sustainability

A third major challenge that wireless networking brands are facing is the need to become more sustainable. Many consumers are concerned about the environmental impact of their wireless connectivity habits, and as a result, they're looking for ways to be greener when it comes to their daily digital lives.

This has created a huge opportunity for wireless networking companies that are able to create products and services that help people reduce their carbon footprint. From helping people reduce their energy consumption by providing energy-efficient devices, or even offering devices that give back to environmental organizations – there are countless examples of how wireless connectivity can be used to help people lead a greener life.

While it's certainly not easy given the wide array of challenges that need to be overcome when creating sustainable products and services, becoming more sustainable is certainly something that many brands are starting to take seriously. Brands are realizing the importance of being able to differentiate themselves from the competition through their environmental impact.

Becoming data-driven

Another important challenge that the wireless networking brands will have to overcome is the need to become more data-driven. Many consumers today are bombarded by so much information every day that they often feel overwhelmed and incapable of making informed decisions when it comes to their wireless connectivity needs.

Brands that are able to provide customers with relevant, actionable information will be able to establish a distinct advantage over competitors that may not be as data-driven. In order for brands to do this, they'll need to invest in the delivery of real-time insights and data analytics that can help them understand how best to serve their customers' specific wireless connectivity needs – and in turn, increase customer satisfaction and loyalty.

5G

Another key challenge relates to 5G, which will be an even bigger game-changer than 4G when it comes to the wireless connectivity industry. 5G will usher in a new age of connectivity, one that is over 10 times faster than current 4G networks.

When 5G becomes truly mainstream, people will be able to do things they've only dreamed about before, such as stream movies and play games on-the-go without delay – and all of this will be possible thanks to the ultra-low latency that 5G offers.

But what makes 5G so exciting from a technology standpoint is also what makes it so challenging for brands that are looking to capitalize on this opportunity. There are still many questions surrounding its rollout and adoption rates, meaning that there's no guarantee that consumers will want to buy into the experience or use their wireless connectivity products and services in the same way as they have with previous generations of networks.

The good news is that once these issues are worked out over time – through increased network coverage as well as customer education about the benefits of ultra-low latency connectivity – the potential for growth in demand for 5G products and services will increase significantly. This means that if you can figure out how to create a compelling user experience for your customers around ultra-low latency connectivity, then you could see tremendous growth as more people become early adopters of this new capability.

To summarize these challenges, brand owners in the wireless ecosystem should consider themselves fortunate: with all these opportunities at hand, there's no reason why any brand can't achieve success by creating products and services that help meet their customers' ever-changing needs.

It's also important to remember that if you can imagine it, then someone somewhere has already created something similar – and now they're trying to take your idea and make it better by leveraging their expertise in hardware design or software development or both.

That's why it's important for brands to think long-term about how they position themselves within an increasingly crowded marketplace. Analyzing product data can help brands overcome these challenges.

Analyzing product data for wireless networking brands

Data has become a critical differentiating factor for brands as consumers seek out products and services that offer them added utility, convenience, or even fun. For instance, much of the early adoption of on-demand transportation services such as Uber and Lyft was driven by their ability to provide riders with real-time updates regarding their location and ETA.

The same can be said for how we use our smartphones – we increasingly expect data-driven experiences from our devices, whether it's instant information about nearby restaurants or attractions while on vacation or seamless integration with household appliances via voice commands or an app interface. And this trend doesn't show signs of slowing down anytime soon – by 2025, the world will be generating nearly 500 exabytes of data a day!

As the number of connected devices increases, so does the amount of data that flows through networks. This increase in traffic is creating a massive opportunity for brands to differentiate themselves by providing users with value-added services and content. Additionally, brands can harness product data, and in particular, product review data, to build more innovative products and services.

Let's take a look at an example of analyzing product review data for a router.

Analyzing wireless networking product review data

A consumer buys a new wireless router, connects it to their home network, and starts using it. The consumer may post a positive review on Amazon about how easy it is to set up the device and use it around the house. This review can help improve product satisfaction and usage among other consumers who read the review. The manufacturer of the router could analyze this product review data to see what features consumers liked most about their device and use that information to improve future products.

On the flip side, a frustrated consumer may post a negative review, which will push away potential buyers. However, this negative review represents a valuable learning experience to find and fix flaws in the router.

By using the data generated from these reviews, it has been shown that manufacturers can improve product satisfaction and usage rates among consumers.

Product feedback is a critical component of any successful product launch or **product lifecycle management** (**PLM**). The PLM process allows manufacturers to understand how their products are being used by customers and generate actionable insights that help them make informed decisions about product design and development. This insight can be used by companies to improve customer experience, differentiate their offerings in the market, and increase sales. The more data that a company generates through its PLM approach, the better position it will be in to achieve success in today's highly competitive market landscape.

Let's look at a practical example of analyzing over 700 anonymized product reviews for a wireless 5G router:

1. First, we'll import our dependencies, or simply the libraries we'll need to analyze the text, sentiment, keyword frequency, and other metrics of the reviews:

```
!pip install wordcloud
from os import path
from PIL import Image
from wordcloud import WordCloud, STOPWORDS
import pandas as pd
import numpy as np
import matplotlib.pyplot as plt
!pip install textblob
from textblob import TextBlob
```

2. Now we can read in the product review data, and turn it all into one string, which we can use to create a word cloud to give some high-level insight into what consumers are talking about:

```
df = pd.read_csv('Reviews.csv')
document = df['Reviews'].to_string()
```

3. We can now use the WordCloud library to generate a word cloud, like so:

```
wordcloud = WordCloud().generate(document)
plt.imshow(wordcloud, interpolation='bilinear')
plt.axis("off")
```

The preceding code generates a word cloud, as seen in *Figure 5.1*, which, besides obvious terms such as router, shows that consumers are talking a lot about **5G** and **range**:

Figure 5.1 – A word cloud of 5G wireless router reviews

We can now use TextBlob to analyze review sentiment and see whether we can find more meaningful insights. We'll want to ensure that each review is properly formatted as a string, and then we can apply the TextBlob sentiment function to each review, or each row in the DataFrame, via a lambda function. This will create two new columns: one showing polarity and one showing subjectivity.

polarity is a float in the range of [-1,1], where 1 means a positive statement, 0 means a neutral statement, and -1 means a negative statement. subjectivity refers to the presence of personal opinion, emotion, or judgment. subjectivity is also a float, which lies in the range of [0,1]:

```
s = df['Reviews']
df['Reviews'] = df['Reviews'].astype(str)
df = df[df['Reviews'] == s]
df[['polarity', 'subjectivity']] = df['Reviews'].apply(lambda
   Text: pd.Series(TextBlob(Text).sentiment))
```

Now that we've calculated sentiment, we can easily sort and search for product reviews by sentiment. For example, the following code will show us negative reviews:

```
df[df['polarity'] < 0]
```

Looking at a snippet of five negative reviews, as seen in *Figure 5.2*, shows us that they're mostly related to the poor range of the router:

	Reviews	polarity	subjectivity
14	Short range and not worth the money	-0.075000	0.200000
21	This router has a limited range	-0.071429	0.142857
27	Range is not very far	-0.038462	0.769231
32	Biggest con - hard to set up	-0.291667	0.541667
57	The router's range is limited	-0.071429	0.142857

Figure 5.2 – A snippet of negative wireless router product reviews

Let's also take a look at more positive reviews, with a positive polarity and fairly low subjectivity, using the following code:

```
df[(df['polarity'] > 0.2) & (df['subjectivity'] < 0.5)]
```

This code will show us snippets of positive reviews, as seen in *Figure 5.3*. We can see that customers are happy about speed, particularly for gaming, but interestingly enough, even positive reviews complain about the range, such as one review that says simply **Good short range**:

163	Fastest router available	0.400	0.4000
179	Best router, good for gaming	0.850	0.4500
185	Good short range	0.350	0.4500
190	Best option for gaming	1.000	0.3000
193	This is the best router!	1.000	0.3000

Figure 5.3 – A snippet of positive wireless router product reviews

These reviews alone show some valuable insights about the router, namely that the consumers like the speed, but complain about the range. Since these reviews are about a 5G wireless router, we can hypothesize that both of these points are related to the nature of 5G, which is incredibly fast, but has notoriously poor range.

This can inform us about a number of possible decisions. For example, the range of 5G routers can be improved simply by adjusting the placement and orientation of the router. Many consumers aren't aware of this, so product teams could use the insights from these reviews to add that information to the product manual, or even change the way the product is photographed to highlight proper router placement.

Additionally, these insights can help inform future product development efforts to prioritize improved range, such as by building more powerful antennas. The wireless networking firm could even build a complementary product, such as a range extender.

Companies today are using virtual assistants such as Siri and Alexa as sources of competitive intelligence – learning about what customers want before launching any new service or offering in order to increase market share and differentiate their offerings from competitors' offerings. For instance, one could analyze how often these digital assistants are used on various networks, potentially highlighting specific features that customers may like and will also use on a network, such as the ability to stream video.

If you're not already doing so, take stock of your company's current level of competitive intelligence-gathering efforts and consider how you might integrate real-time customer data into your existing processes – this will give you even more insight into how customers are interacting with your brand today (and perhaps tomorrow).

Using Commerce.AI for wireless networking brands

The wireless networking industry is one of the most mature, yet dynamic, sectors today. From smartphones to IoT devices, and from smart homes to self-driving cars, everyone relies on wireless technology for communication and connectivity.

Wireless networking brands are in the business of connecting people – and that's a big opportunity. Consumers want to use their favorite brands across multiple platforms – *so how can connectivity brands create value for customers? How can they capitalize on this massive opportunity?* In the following sections, we'll explore how brands can answer these questions with data-driven solutions.

Enter data-driven solutions

The role of data-driven solutions in the wireless networking industry has evolved tremendously since 2020. At the turn of the century, only a tiny minority of consumers had access to online retail sites. Today, over 90% of all US consumers have made a purchase online (https://optinmonster.com/online-shopping-statistics/#:~:text=That's%2091%25%20of%20the%20country's,only%20ones%20who%20shop%20online).

With so much consumer demand for connectivity products and services, brands must leverage data analytics if they want to stay competitive.

Brands can do this by harnessing machine learning algorithms and complex math models in order to uncover patterns in large datasets where humans might not see them otherwise. They work behind the scenes, crunching numbers that help businesses make smarter decisions that then lead to tangible outcomes such as new product launches or improved customer engagement rates.

Let's look at how brands can use Commerce.AI's machine learning algorithms and data engine to improve their **key performance indicators** (**KPIs**). Commerce.AI has built the world's largest product data engine, which has analyzed over a trillion data points from sources such as Amazon, Walmart, Target, and even YouTube video reviews and voice surveys.

Prediction is the ability to infer trends from data that may or may not be reflected in present trends. The more data you have, the better your prediction capabilities. In this way, a data engine can help acquire and manage product data in a more strategic and proactive manner that can help increase profitability.

In particular, brands can implement machine learning models to predict and optimize a number of important KPIs, as shown in *Figure 5.4*:

Key Performance Indicators (KPIs)

✓ Maintain or increase star rating trend   ✓ Improve conversion

✓ Improve best sellers ranking   ✓ Improve search result ranking

✓ Decrease time compiling weekly reports   ✓ Improve detail page glance views

✓ Improve product sentiment

Figure 5.4 – Important KPIs for product teams

Let's look at each of these KPIs in detail to understand why they're important for product teams to consider, analyze, and optimize.

Star ratings

The importance of maintaining or increasing customer satisfaction ratings cannot be overstated. Brands that fail to do so are at risk of losing customers and, ultimately, revenue.

Today, more than ever, consumers want to know what they're getting for their money. For many people, the most important aspect of a purchase is the value received in return for their hard-earned cash. This means that brands need to ensure that each and every transaction with a customer is a positive experience – and not just when it comes time to make a purchase but also throughout the entire customer life cycle.

This starts with ensuring that product quality and service meet or exceed customer expectations when making initial contact with the brand and continuing through all stages of the relationship with customers (for example, from initial contact through repeat business).

Remaining competitive in today's highly saturated marketplace requires strong focus on acquiring new customers while retaining existing ones too; our analysis indicates that negative feedback posted online by unhappy customers can have serious implications for companies' bottom lines (*in other words: don't ignore negative reviews!*).

Social media platforms such as Facebook and Twitter have made it easy for people to voice concerns online about products or services they have purchased. This allows them to voice any concerns they may have regarding purchases made through competing vendors (or even past purchases from the same vendor).

Importantly though, it also allows them to provide feedback directly related to these previous experiences too – such as poor product quality or shoddy customer service delivery when compared against similar products/services offered by competing vendors/brands.

And given how heavily relied upon consumer recommendations are today when making purchasing decisions, negative online reviews can prove decisive when it comes time for consumers to make purchase decisions.

With Commerce.AI, product teams can analyze reviews, the reasons behind reviews, and trends in reviews across virtually all product sources, helping to inform new product innovation, marketing, and customer support efforts.

Improving best sellers ranking

Best sellers ranking is simply a list of the top-selling products on a given site. It is an important metric for retailers because it indicates which products are popular and which ones to stock more of or sell out faster when they're running low.

In other words, it indicates which products retailers should be selling more of or focusing on driving higher sales for. In fact, many retailers have dedicated resources to manage their best seller's strategy – and even create new best sellers – as part of their overall business strategy and operations teams.

So how does one determine what's working and what isn't when it comes to driving sales? By analyzing sales data over time – both historical and real-time – and identifying trends that can be leveraged to improve performance in the future.

Historical analysis helps define patterns by looking at previous periods (for example, last year, last quarter) when certain events occurred (for example, an event occurred such as an economic recession). This provides valuable insight into what worked before – and often provides clues about what may work going forward too.

Real-time analysis looks at factors such as clickstream data, visitor behavior, and mouseover actions taken, along with other metrics, such as conversion rates from sign-ups through email campaigns or website forms.

Simply put, analytics – both historical and real-time – helps you identify trends that can be used to inform your marketing efforts going forward so you can drive more conversions, higher sales, increased revenue, and greater profitability with your product portfolio.

Time compiling weekly reports

Leading wireless network brands have approached Commerce.AI for help with their reporting process. Since large wireless networking brands have more than 100 products and services across multiple verticals, they find it difficult to stay on top of what's going on across all of their initiatives and businesses – and even within their own departments.

One reason for this was that they'd been using spreadsheets and word processing programs to create their detailed KPI reports, which meant that they had to hire additional team members just to maintain these reports – a costly endeavor.

With so many products and services under different management groups across multiple verticals, it also proves difficult to share and communicate information about what is happening in various areas.

By automating reporting using AI-powered tools, our clients are able to eliminate the need for dedicated staff just to maintain spreadsheets. Via dynamic visualizations that show patterns in performance across multiple products and services from a variety of angles, clients can spot trends much earlier than before (plus it saves money on HR costs).

This enables clients not only to take action in real time but also to identify trends earlier than ever before so that they can proactively manage risk and make informed decisions about growth opportunities rather than reacting after something has gone wrong.

By analyzing millions of pieces of data every day from thousands of sources around the world, we are able to give clients actionable insights into where problems might occur, or opportunities might present themselves.

Improving product sentiment

Sentiment analysis is a form of machine learning that can determine the overall tone and sentiment of a piece of text or an entire product's reviews. It's used to understand what people are saying about brands, products, and services on social media – and then take action accordingly.

For example, if customers are expressing positive sentiments about a particular product or service on social media, it could be an indicator that they will likely buy this product in-store as well (if the brand has a brick-and-mortar presence). On the other hand, negative sentiments could indicate that the brand should consider improving its product or service offering.

In the past, wireless networking brands would have relied on qualitative customer experience research (such as in-person interviews), but now they have access to a wealth of data – and it's all about people's online experience with their products and services.

By using AI technology, these brands are now able to gain actionable insights from their massive amounts of product sentiment data that were previously unavailable. Let's take a closer look at how this works.

Machine learning

The data engine processes vast amounts of unstructured text data (such as reviews and comments) to identify patterns and trends. It then uses those insights to inform future decisions.

For example, based on the latest reviews posted about a particular product or service, the AI engine can determine which aspects need improvement and then suggest ways for manufacturers to improve them – such as improving the speed of shipping or developing new features in the product itself. *This is machine learning in action!*

Data mining

Before AI could be used effectively by wireless networking brands, they needed to be able to process massive amounts of text data (that is, review posts). They also needed tools that could mine their vast stores of such information quickly and efficiently so that they could spot trends and perform sentiment analysis on an ongoing basis.

That's where Commerce.AI comes in. We provide them with a cloud-based platform that allows them to easily manage large volumes of textual information – and provides real-time insight into what customers are saying about their products and services on social media platforms such as Twitter, Facebook, Instagram, and so on.

So far this year, we've seen some impressive growth from our wireless networking vertical clients, who are using AI technology in ways that were not possible even just a couple of years ago, given the tremendous growth in wireless networking data amid the COVID-19 pandemic.

Actionable insights

Once wireless networking brands have enough meaningful insights from their product sentiment analysis, they can use this information to improve their products or services accordingly – and make sure that customers are happy when it comes time for repeat purchases. This is another important benefit of using AI technology for product sentiment analysis – it helps manufacturers create more meaningful customer experiences than ever before.

Improving product conversion

A product conversion rate measures the percentage of visitors who make a purchase after visiting a website. The more traffic you get to your e-commerce website, the higher your product conversion rate will be. This metric is important for e-commerce retailers, because it shows how well they are converting visitors into customers.

By looking at site traffic and purchase metrics, we are able to figure out what factors correlated with higher or lower conversions on product sites, and highlight the following:

- Product categories where there was high engagement

- Product subcategories where people spent more time researching before buying (for example, desktop computers)

- Product pricing pages where people had conversations about pricing strategy before making a purchase decision (for example, laptops over $700).

- Product reviews where people compared prices between competing products and then made their purchase decision (for example, laptops under $500)

Search result ranking

In a nutshell, **search result ranking** is based on an algorithm that determines the order in which results are displayed when a user performs a search on a website.

The importance of search result ranking cannot be overstated. In fact, it's one of the most important factors determining whether or not a consumer will click on an ad or purchase a product. Consumers are more likely to convert on high-ranking results than organic or paid products from companies that appear lower in their searches.

This is because people tend to trust sources they see higher up in search results. So clearly, there's value for brands in optimizing their position in search results. *But how can data and AI help?*

The data engine behind Commerce.AI compiles a wealth of information on brands, including their performance in search results over time. This enables us to predict the likelihood that a brand will appear higher or lower in search results, as well as its overall ranking.

Analyzing data from previous searches helps wireless networking brands optimize their position based on actual performance and not just guesswork or hunches. In other words, it allows them to improve their chances of appearing higher in search results by investing in activities (such as content marketing) that help improve brand affinity and increase **click-through rates** (CTR).

By using data from previous searches, they can also identify patterns and trends that give them insight into what works and what doesn't work when optimizing for search result ranking. For example, if traffic to a website is low but the CTR for ads on that site is high, there's a good chance the company has done something right with regards to optimizing for CTR.

On the other hand, if traffic to a website is high but CTR for ads on that site is low, there's probably an issue with ad copy or relevance of product listings that should be addressed before moving forward with any investment in improving CTR further (that is, by ensuring relevant product listings are included).

By identifying such issues at an early stage rather than waiting until things go wrong (that is, after spending money on increased visibility), wireless networking brands can save themselves time and money while still achieving their desired outcomes – higher rankings – without compromising their brand values or integrity (in terms of ad copy/relevance).

Detail page glance views

Let's start by defining what we mean by a detail page glance view: the number of times an Amazon product detail page is viewed during a defined period (for example, 1 day). This metric has become increasingly important as it reveals patterns in how users interact with products on Amazon – and where there are opportunities for growth.

In the case of wireless networking brands such as TP-Link, Netgear, and Linksys that sell Wi-Fi routers and related hardware, the ability to track these metrics provides insights into how consumers are interacting with their products on Amazon. It can also provide valuable information about which elements of a product detail page should be optimized in order to improve CTR, conversion rates, and overall sales volume.

The importance of this data cannot be overstated for any brand or business that sells goods or services online via e-commerce marketplaces such as Amazon. In fact, it's so critical that many major brands have dedicated entire teams to monitor it. In our experience working with multiple brands on online retail optimization, having visibility into these metrics has been critical in helping them determine which elements of their products they should prioritize when creating new versions of those items for sale on Amazon.

It goes without saying that no retailer wants to spend time and resources creating new versions of an item only for it to fall short in the market. Those problems could have been easily avoided had retailers been able to track how many times consumers viewed each version during a given period (that is, glance view data).

Summary

In this chapter, we've learned about the key challenges of wireless networking brands, including 5G rollout, deploying sustainability measures, increasing complexity, and so on. We've also learned how to analyze product data and use Commerce.AI to overcome these challenges and build better wireless networking products and services.

AI has the potential to help wireless networking brands overcome key product-related challenges, including meeting more demanding user needs and standing out amid an increasingly competitive market.

In a traditional product intelligence environment, analysts must manually sort through copious amounts of data to find market opportunities, analyze competitors, and create innovative new products.

With an AI solution, billions of products data points can be analyzed effortlessly to create competitive advantage in the wireless networking market. By understanding challenges and AI-based solutions in the wireless networking market, you can begin to see how AI can be useful for any product team.

In the next chapter, we'll explore how to apply AI for innovation in the consumer electronics space. As with wireless networking brands, consumer electronics firms face a unique set of product-related challenges, which can be addressed with AI due to its versatile ability to learn from large amounts of product data.

6
Applying AI for Innovation – Consumer Electronics Deep Dive

As we've explored in previous chapters, AI is no longer just a buzzword. It has become a critical component for the growth strategies of many companies, with the majority of leading executives saying that their company is investing in AI or machine learning. This chapter will explore how consumer electronics brands can leverage AI to improve their product innovation and drive growth.

Consumer electronics brands have long relied on their ability to innovate new products to keep pace with the rapidly changing trends in technology. Innovating new products enables these brands to stay relevant and attract consumers who want to experience new ways of engaging with technology.

This applies as much today as it ever has. For example, when consumers are choosing from an array of smart home devices at their fingertips, they need compelling reasons to choose your brand over someone else's. By taking advantage of emerging AI technologies, consumer electronics brands can create products that are more immersive, interactive, and enjoyable than ever before – enabling them to stand out from the increasingly crowded tech shelf.

In this chapter, we'll cover the following topics:

- Understanding the challenges faced by consumer electronics brands
- Analyzing the product data for consumer electronics brands
- Using Commerce.AI for consumer electronics brands

We'll learn about how consumer electronics brands are facing new challenges when it comes to the connected consumer, the content consumer, and greater competition from all sides. We'll also explore how to collect, analyze, and use consumer electronics data to become more innovative and overcome various challenges.

Understanding the challenges faced by consumer electronics brands

Let's start by exploring the challenges of consumer electronics brands to understand why new, innovative, data-driven, and AI-based solutions are needed to drive product success.

Some of the challenges we'll cover include the needs of the connected consumer, the new reality of short-term attention span, the demands of the content consumer, and growing competition from emerging markets.

The needs of the connected consumer

The **connected consumer** is a recent phenomenon, but one that has become the new normal for consumer electronics. Practically, every consumer now expects to be able to connect to the internet and engage with their technology in some way.

This isn't just about smartphones and tablets anymore. It's about wearables, home automation, and smart speakers such as Alexa or Google Home. All of these devices enable consumers to interact with technology in meaningful ways.

Brands have been slow to recognize this shift as it's still relatively new. But that will change over time as more people adopt these devices for everyday use, which is why brands need to pay close attention now if they want to optimize their business strategy moving forward.

For brands to succeed in today's connected world, they need a strong identity – and that starts with understanding how people want to interact with them through technology. To do this effectively, brands should look at how people currently use technology (in other words, *how are you already winning?*).

Then, brands need to think creatively about how they can leverage those strengths into something bigger – something that provides them with more opportunities for engagement and deeper relationships with their customers.

For example, if we think back to when the first smartphone came out – the iPhone – it changed everything because it was incredibly accessible and intuitive. Apple took advantage of this familiarity and made the switch to touchscreens much easier than any other company could have done. It did this by leveraging its strength in product design as well as software development.

The importance of product design cannot be overemphasized here: people don't merely buy things; they use things and experience things. If your customer doesn't naturally gravitate toward using your product because it feels natural or intuitive, then neither will the market at large.

A new reality of short-term attention span

The average attention span used to be far longer. Today, it's commonly said to be just 8 seconds. People have many options for what they want to pay attention to and what they don't want to pay attention to – and with social media platforms from Twitter to TikTok, there is no shortage of opinions about brands and products.

The impact on social media

Brands must think about their digital footprint as much as their physical footprint if they are going to succeed in today's increasingly digital world. In other words, how consumers perceive a brand on social media matters just as much as how they perceive it in person when making purchasing decisions.

Today, more than ever before, consumers are aware of the power they wield over brands through social media – even if some brands might not always act in a way that respects those powers (and can get them into hot water).

For any brand – whether it's a consumer electronics brand or any other kind of brand – to be successful today, it needs an active and engaged community online that feels connected to the brand and its values. This means actively engaging with your community on various channels (for example, Facebook groups, Instagram stories, and TikTok shorts) so that you can build meaningful relationships with them.

The key here is figuring out how best to use your existing resources (your employees) while at the same time identifying ways you can add new resources (such as contractors and freelancers) that will help take your company's marketing efforts to the next level.

The impact on product teams

It's not just the social media landscape that's shifting due to today's digital first world. A short attention span means that consumers are less likely to be willing (or able) to invest the time and energy into fully experiencing a product that they have little or no emotional connection with.

To maximize engagement with their content – whether it's a movie trailer, song lyric video, TV show preview, or book excerpt – the consumer electronics brands need to think about how they can create more engaging experiences that go beyond just the content itself.

This requires understanding how to use technology in ways that enhance the consumer experience of your content instead of replacing it. This includes allowing users to interact with your content in new ways (such as AR glasses for interactive storybooks); using movement-based media (such as dancing GIFs) instead of static images; creating *cinematic* experiences through 360° videos; using live streaming platforms leveraging virtual reality platforms; and so much more!

The bottom line is that the consumers are spending less time engaging with traditional media and more time engaging with digital media – and brands should utilize this opportunity to build meaningful relationships with their communities.

Meeting the demands of the content consumer

The content consumer has changed since the days of the compact disc. Today, consumers are bombarded with a constant stream of information and entertainment. The rise of online video services such as Netflix and YouTube have made it easier than ever for consumers to access an immense amount of content at their fingertips. *So, how does a consumer electronics brand differentiate itself amid this sea of video content?*

The answer is by creating compelling experiences around the content itself, rather than trying to compete on the price alone. People watch billions of hours of videos each month. As such, brands must think beyond just the product itself when delivering compelling experiences around their offerings. They must create compelling narratives around their products and services to stand out in a crowded marketplace where there is plenty of competition for customers' attention and wallets.

Content creation is no longer the domain solely for large media companies or Hollywood studios; smaller creators can now reach massive audiences with relatively little investment in production value or infrastructure.

What this means for brands is that they can't simply rely on their product offering as a way to differentiate themselves from competitors; they need to go above and beyond with engaging content if they want to capture customers' attention and loyalty in an increasingly competitive marketplace.

Brands should also pay close attention to customer expectations regarding social media engagement throughout the entire customer journey – from product development through distribution and beyond – so that the customers feel included throughout each stage of the experience process.

The need to become data-driven

Consumer electronics brands need to become data-driven. Data is the new *oil*. It's what fuels a company's growth and innovation and it's what makes or breaks consumer brands.

In many cases, consumer insights come from data analysis, not just human intuition. The companies that can leverage data effectively will be able to improve product and service offerings while increasing engagement with the customers.

On the other hand, those that don't have robust analytics capabilities will be at a disadvantage compared to their competitors, who do. *So, how do companies go about becoming more data-driven?* We'll answer this in the *Analyzing product data for consumer electronics brands* section.

Emerging consumer electronics markets

The rise in the popularity of crowdfunding and the explosion of interest in new and innovative technologies have created a boom for the consumer electronics industry. The number of people who use technology daily is in the several billion and is continuing to grow rapidly in emerging markets. Some see this as an opportunity for disruption, while others view it as a threat.

Some experts believe that with so many new users entering the market, incumbents will be challenged to maintain their dominance due to increased competition.

Brands must also contend with rapidly changing trends, such as **virtual reality** (**VR**). Today, we are already seeing major brands such as Apple, Samsung, and Facebook enter the VR space. As consumers become more familiar with VR technology, it's likely that traditional brands will adjust their strategies accordingly – and could even get bought out by the tech companies.

To stay competitive in this environment, you'll need to think about how you can attract more customers while maintaining your brand identity across multiple channels – digital and physical retail locations, online stores, and e-commerce platforms are just some examples of where you can engage your audience today.

Now that we understand some of the major challenges faced by consumer electronics brands, let's look at how to analyze the product data to overcome these obstacles and move ahead.

Analyzing product data for consumer electronics brands

Consumer electronics firms rely on product data to understand their customers and market trends. Product data is critical for understanding how your products are performing, what consumers want, and how they interact with your brand.

The data you collect can help you identify issues, measure success, and make better strategic decisions. But it can be difficult to find the right kind of product information that's actionable and helpful – especially if you're not a consumer electronics expert.

The data-driven product strategy is about using data and analytics to develop new product ideas, evaluate existing products, and improve the overall experience of customers. The goal is to create more customer engagement by changing how people interact with your brand.

The data-driven product strategy has many benefits, including the following:

- Increasing customer loyalty and retention through targeted offers
- Lowering costs while maintaining quality by streamlining operations
- Increasing profitability by improving revenue or increasing margins through innovation
- Reducing risk in the marketplace through enhanced understanding of consumers' preferences

Key considerations in the data-driven product strategy

Each company's data landscape will be different, depending on its size, industry structure, and business model. At its most basic level, a data strategy involves creating a dataset that can be analyzed for insights into consumer behavior.

For example, you might collect behavioral data from customers who have recently purchased one of your products or services. You could also look at historical sales to identify patterns that indicate which combinations of features are the most popular among customers. This information can then be used to inform future product development and marketing campaigns.

Companies can also use AI and machine learning algorithms to determine what types of products their customers are likely to buy based on their behavior online or offline, for example, by looking at past purchases or search history. Finally, companies can analyze this information with teams across various departments (marketing, engineering, and finance) to identify opportunities for new products or services that meet the needs of the consumer within the broader community of users they serve.

In our experience, we can advise the clients to build a data strategy for consumer brands across all industries e-commerce, FMCG, and B2B). Here are some key considerations:

- **Focus on customer needs**: Identify what you know about your customers' behavior based on the existing datasets or market research studies (online/offline). Then, identify what you don't know about them yet – *what do you need more data points for before you can start validating hypotheses?*

- **Define roles and responsibilities**: *Who will own each stage of the process? How long will it take? What tools/methods will they use?* Defining these roles is crucial to ensure that the project gets done and has ownership.

- **Build capacity**: *Who will ensure that there is an ongoing commitment from senior management?* This will ensure that oversight and support are provided when necessary.

With Commerce.AI's data engine, most (if not all) of the data you need will already be available. Our product's data engine features over a trillion data points on hundreds of thousands of products and services, retrieved from over a hundred sources in a variety of languages.

Now that we understand the key considerations in the data-driven product strategy, let's look at how to collect that data.

How to collect consumer data

Consumer data is the lifeblood of any brand. A strong consumer data program can help a company understand what motivates the customers to buy and, ultimately, influence their purchasing decisions.

In the consumer tech world, product reviews are an accepted and effective way for brands to collect customer feedback on products. A review from a satisfied customer can be just as influential as a glowing press release in marketing or PR campaigns.

Another great source of customer feedback is social media: Instagram posts can provide valuable insight into how the consumers feel about your product's aesthetic appeal and ease of use, while Twitter feeds can show how people interact with your brand through jokes, memes, and other forms of social commentary.

You should also pay close attention to what people are saying about your competitors' products – a competitor's post about new features in an upcoming update might catch fire among existing users who want to stay up to date on the latest features before making their next purchase decision.

With so much consumer data available at our fingertips these days (through online forums such as Reddit, app analytics platforms such as Mixpanel, and email marketing services such as Mailchimp), it can be easy to overlook some less obvious sources of product feedback that are often more actionable than written reviews or tweets.

Speaking candidly about experiences with problems with your products is one such source of valuable input that many companies fail to capitalize on. This is due to a lack of experience collecting this type of feedback directly from customers. This is where Commerce.AI voice surveys come into play: they enable businesses across all industries to listen in real time as people voice their concerns about their products through audio.

By taking advantage of tools like these – combined with robust online communities such as Reddit – brands can learn more about what's driving customer satisfaction or dissatisfaction than they ever could, by simply reading user reviews online.

How to integrate data into the product design

A lot of research has been done on how to design products that are more appealing to users, but very little research has been done on how to design products with the end user in mind. There are four stages in the end-to-end persuasive design process:

1. Understanding your users
2. Using personas

3. Creating personas based on data

4. Identifying the pain points within each persona

Understanding your users

Understanding who you're designing for is important because it helps you understand what they want from a product and why they might want it. The more you know about your users, the better designed your product will be (and the less likely it will have usability issues).

In the past, product designers had to rely on intuition and anecdotal evidence to understand how their users used the technology. Nowadays, with the availability of large amounts of data, it is possible to learn a lot about the users by observing how they interact with products.

For example, you can observe how long each user spends on a screen or a page before clicking away, or which parts of your site are least used. You can also see where people click or scroll before making a purchase decision. All this information will help you improve the usability of your product and ultimately increase its conversion rate (the percentage of visitors who become customers).

Tracking data also allows us to iterate quickly on the design until it matches our ideal user experience, by continuously running **A/B tests**.

Using personas

Personas are a tool to help you understand your users. They can help you identify needs, goals, and motivations that will provide information about the design of your product. A persona is a fictional representation of an individual that represents some aspect of a target user group. The idea is to create a living document that describes the target user's characteristics, motivations, and expectations.

This allows you to test the assumptions about how your product or service might be used by real people before building it. Personas are useful because they allow you to focus on what's important, instead of assuming how users will behave. It also helps prevent making the same mistakes over and over again with new users by focusing on who those new users are, rather than trying to guess how they might behave based on past trends or the behaviors of other groups in your target market.

You can use personas in many different ways as part-of-the-puzzle pieces, including identifying pain points in the design process, understanding customer needs and wants early on during development, creating an internal vision board for your team, or even testing new ideas with the potential customers before building something just because it looks cool.

Creating personas based on data

You now have a list of people who fit certain characteristics and motivations, but this isn't enough information on its own – *you need data too!* If possible, try to find quantitative ways of understanding who these people are and what motivates them. By doing this, this information can be incorporated into your personas, as well as into the rest of your analysis process.

For example, instead of creating fictional characters based on qualitative observations, create an analytics report showing those observations numerically. This information could then be used by someone at an e-commerce company to estimate the sales volume they should expect if they ran an online store during the Black Friday weekend – something qualitative insights alone wouldn't necessarily tell them.

Identifying the pain points within each persona

Now that we have our personas figured out, the first step is finding where they don't get everything that they want, need right now, or easily enough – these are their pain points.

A **pain point** is a problem that a customer experiences with a product or service. It starts with identifying where your customers are struggling with your product and then finding ways to provide them with the solution they need.

At its most basic, this process should start with asking yourself, *what don't my customers have?* This is important because it forces you to think about what features you want to build into your product and how you can make those features more accessible.

Once you have identified these pain points, the next step is determining which ones are critical enough for you to invest in solving for your customers. Remember, failing to solve an important pain point will mean that users will simply look elsewhere for what they need, so you must pick the right ones.

It's important to remember that not every pain point needs to be solved by building new products or services. Sometimes, there are existing solutions on the market already, and sometimes, there are things that can be done within the existing constraints of technology or design.

To figure out which problems need solving, we often look at existing trends and patterns in our industry, as well as data from similar companies within our space that have been successful at addressing these problems before us. We also ask ourselves questions such as the following:

- *Do our current customers feel like they're missing something?*
- *Is there another way of doing this already in existence?*
- *What other companies within our space do well?*
- *Do we have anything unique, but can we borrow from their playbook?*
- *Are there any patterns we see across industries when looking at similar products/services?*

The answers to these questions help inform you of what pain points to address, which is a crucial component of product innovation.

Now that we know how to collect, analyze, and use consumer electronics data, let's explore the next stage of turning this data into insight: using Commerce.AI.

Using Commerce.AI for consumer electronics brands

As Commerce.AI runs the world's largest product data engine, there's a whole host of opportunities for consumer electronics brands to become more innovative. Let's explore how to use Commerce.AI to better understand product positioning, analyze the consumer electronics market, improve research initiatives, generate product ideas, and more.

Understanding product positioning

Understanding product positioning is critical for understanding the market potential for any given product. The following diagram shows how the Commerce.AI positioning chart can be used to understand where the different products stand in their market space by comparing their sentiment scores (vertical axis) and **Number of reviews** (horizontal axis):

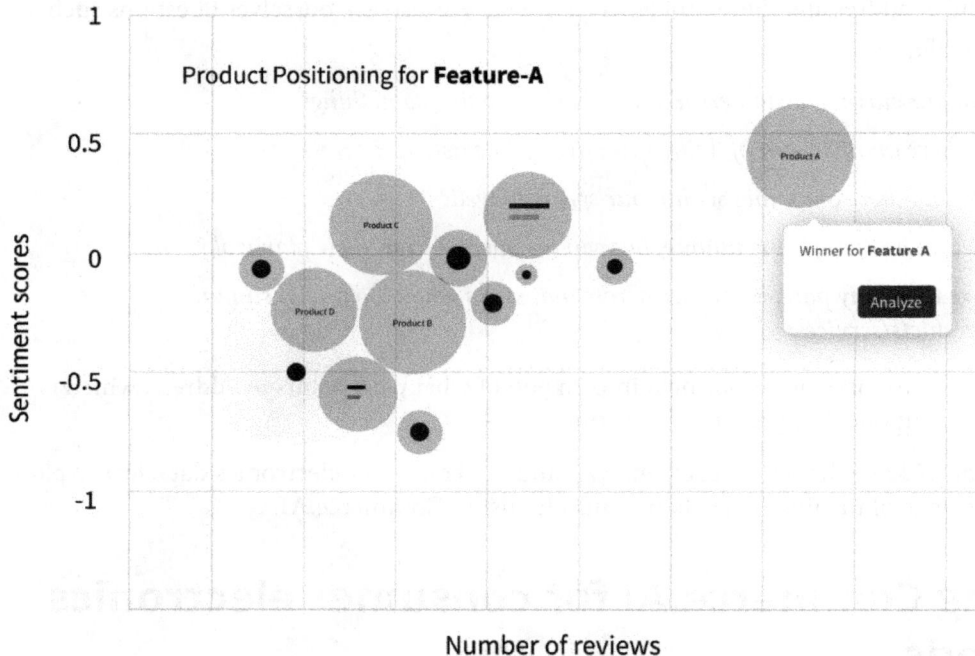

Figure 6.1 – A product positioning chart mockup comparing sentiment and number of reviews

In the case of a camera, for instance, an entry-level point-and-shoot model may have a lower position on the chart than a higher-end digital SLR that has better features such as a faster lens, a larger sensor, and more advanced image processing capabilities.

By comparing these products across multiple dimensions of customer value, you can quickly gain insights into the relative value of each product in its market space. By looking at only one dimension (for example, sentiment), you would miss out on important insights. For example, if a certain camera has a very high sentiment, but only 10 reviews, then that sentiment score may not stand up to scrutiny and may fall apart in the real world.

Analyzing the market with consumer electronics AI reports

The consumer electronics market is huge. It's forecasted to surpass $1.2 trillion in sales by 2022, according to Statista research (`https://www.statista.com/markets/418/topic/485/consumer-electronics/#overview`). Besides smartphones and laptops, countless other devices fit into this space, from tablets to VR headsets to home automation devices. Depending on your definition of the term, there may even be room for wearables such as fitness trackers or smartwatches, if you're willing to stretch the definition a bit.

With so much money being spent on these gadgets, it's no surprise that brands want to know what people are buying and why they're choosing these products over others. Market research firms conduct extensive research into consumer behavior and preferences to understand how trends will play out in the future, as well as what needs people have that can be fulfilled by brands through their products and services.

The problem is that this enormous market is a double-edged sword for product teams: There's a tremendous financial opportunity at hand, but with so much data, it can be difficult to make sense of things. It'd be impossible to manually analyze the product sentiment and reviews of the millions of products out there.

With Commerce.AI's AI-generated market reports, this data is autonomously analyzed to provide insights at unparalleled speeds. What would have previously taken teams of researchers months can now be done in a few clicks.

The following screenshot shows an AI-generated market report on **DSLR Cameras**, including the number of products in the relevant **Amazon** category, the fastest-growing brands, the best-seller products, the number of **Brands**, and an **OPPORTUNITY METER** that summarizes the size of the market opportunity:

AI Generated Summary

The DSLR Camera segment has an opportunity level of 54 with 1,992 products in the category. Amazon reported a steady increase in sales of consumer electronics whereas DSLR Cameras has a growing demand.

Fast Growing Brands

Canon, Nikon, Panasonic, Loupedeck, Fujifilm, Sony, Deal-Expo

Bestseller

Nikon D3500 W/ AF-P DX NIKKOR 18-55mm f/3.5-5.6G VR Black (**4.8 stars, 2,991 ratings**)

OPPORTUNITY METER

41-54%

BRANDS

7+

Figure 6.2 – A snippet of an AI-generated market report on DSLR cameras

Brands need to understand how the consumers are using their products; it provides valuable insights into potential product improvements or new features that can be added before the competitors get there first. Market research also allows companies to maintain a competitive advantage by staying ahead of trends before they become mainstream consumer behaviors.

How does Commerce.AI help with consumer electronics brand research?

One way for brands to gain insight into consumer behavior is through analytics software designed specifically for market research purposes.

Since Commerce.AI operates the world's largest product data engine, with over a trillion data points analyzed, the data that's presented to a brand is carefully curated to their needs. Otherwise, there would be information overload, which is the status quo that product AI aims to break through with focused insights. The following diagram shows how a **Consumer Electronics** dashboard in Commerce.AI is like a *blank slate*, which fills with the relevant brand data:

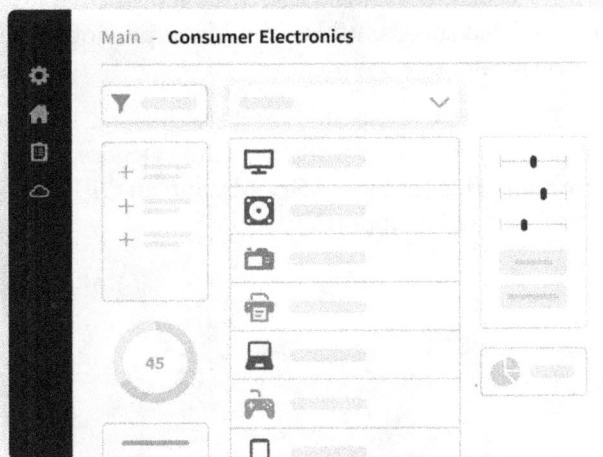

Figure 6.3 – A mockup of the blank slate Commerce.AI dashboard

Consumer electronics brand research requires analyzing an enormous amount of data. That's where Commerce AI comes in – we provide powerful analytics software designed specifically for use within the commerce businesses.

Our platform was built from the ground up; we understand the types of data you need at your fingertips when managing a product line. And now, we're taking our expertise one step further by adding deep learning capabilities so that you can gain actionable insights into the consumer behavior inside your product and service line.

Generating consumer electronics product ideas

In addition to analyzing the market and researching trends, we can even use AI to generate consumer electronics product ideas from scratch. **Product ideation** is a key component of innovation, as failing to come up with novel and exciting ideas can ultimately lead to business failure. After all, the most successful companies today, from Apple to Tesla, are the ones that thought outside the box and broke the status quo.

Let's explore an example in Commerce.AI that uses large language models to generate product ideas. These models are used to predict the most likely words in a text, given its content, similar to the way that the traditional **natural language processing** (NLP) technologies can be used to analyze text, but at a much larger scale.

Large language models have recently become more widely available thanks to improvements with architectures such as the **Transformer**, which are now fed on massive amounts of textual data, which includes product descriptions and reviews from all over the internet.

Large language models are particularly useful for generating product ideas because they learn from natural human language rather than from individual words or phrases. As such, they consider broad concepts such as *computer* or *tablet* as well as more specific concepts such as *iPad* or *iPhone*.

The following screenshot shows how this is done in practice. First, you'll enter a **Product Category**, such as DSLR Cameras. Then, you can optionally select a **Customer Wishlist**, which is extracted from product reviews. This **Customer Wishlist** will be used to inspire the generated ideas. If no wishlist is selected, then the review data will be used to holistically generate ideas. Another area to note is **Creativity**, which is a type of randomness setting for the large language model. High creativity or high randomness will result in more out-of-the-box, but perhaps less grounded and realistic, ideas:

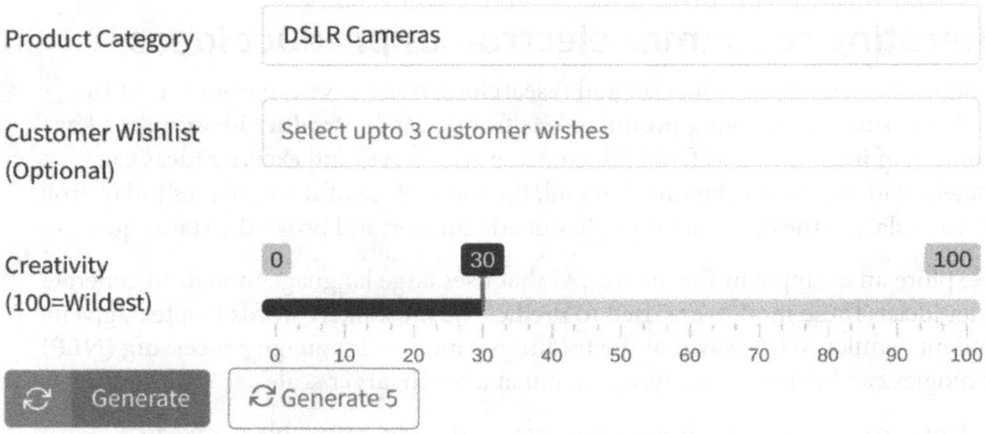

Product Category	DSLR Cameras
Customer Wishlist (Optional)	Select upto 3 customer wishes

Creativity (100=Wildest)

0   30   100

0   10   20   30   40   50   60   70   80   90   100

⟳ Generate   ⟳ Generate 5

Figure 6.4 – A product positioning chart mockup comparing sentiment and number of reviews

Using this approach, we've been able to generate thousands of unique products with little effort. In the following screenshot, we generated ideas for new DSLR product ideas, such as a DSLR that you can control from your phone, a DSLR that responds to voice commands, and a DSLR that processes complex colors accurately by using AI:

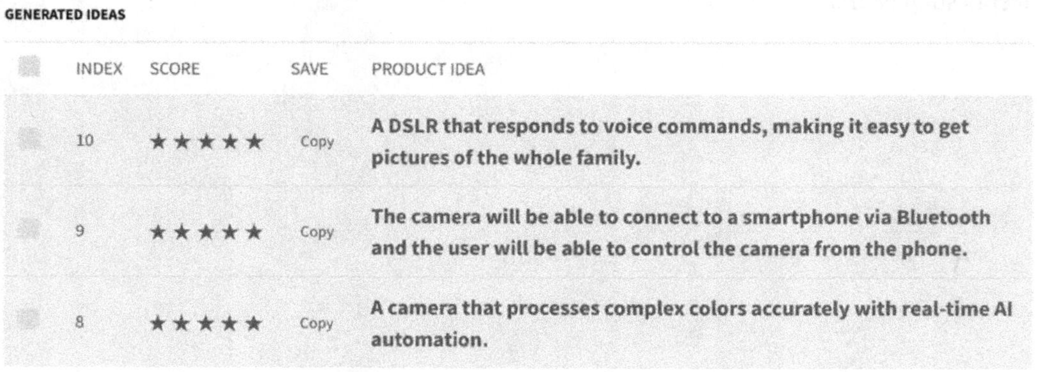

GENERATED IDEAS

	INDEX	SCORE	SAVE	PRODUCT IDEA
	10	★ ★ ★ ★ ★	Copy	A DSLR that responds to voice commands, making it easy to get pictures of the whole family.
	9	★ ★ ★ ★ ★	Copy	The camera will be able to connect to a smartphone via Bluetooth and the user will be able to control the camera from the phone.
	8	★ ★ ★ ★ ★	Copy	A camera that processes complex colors accurately with real-time AI automation.

Figure 6.5 – A snippet of AI-generated DSLR product ideas

While AI won't replace human ingenuity, it can help augment our creativity. Practically, all product teams have experienced the feeling of *creative block*, in which it's difficult to come up with new ideas. By using AI, consumer electronics product teams can get those creative gears turning and perhaps help find their next big idea.

Extracting insights from Shopify

Shopify is a platform largely aimed at **small and medium-sized businesses (SMBs)** to power their online stores, including inventory management, shipping, payments processing, marketing tools such as content creation and analytics, and more. Since its launch in 2004, Shopify has grown into the world's largest independent multi-channel retailer.

Commerce.AI can be used to extract insights from your Shopify store by sending voice surveys directly to your Shopify customers. Voice surveys have higher engagement and completion rates than their text-based counterparts, as many people find it easier to speak freely than having to sit down and write their thoughts out.

The data you receive from these surveys can be used to inform product development decisions. For example, your data might show that certain aspects of a product or service are particularly important to your customers. This knowledge can help you make informed decisions about the types of products and services you should offer to satisfy your end users.

As consumers, we all have our individual preferences when it comes to shopping online. Some people prefer to read product descriptions before making a purchase, while others prefer to look at the images and video content instead. As an e-commerce brand, you should consider your target customers and what they find most valuable while browsing your website; this will inform the e-commerce brands about how you design your store and how you format your product descriptions and imagery.

Additionally, some people prefer to interact with consumer electronics through voice rather than text; this is because it's easier for them to speak naturally without having to worry about typos or spelling mistakes.

With over a million merchants using Shopify, there are plenty of opportunities to better understand the market through voice surveys within your customer base – all without any additional development or staff time investment required.

Sharing insights on Slack

Innovation success is largely dependent on successful communication. The ability to effectively communicate new product concepts, and the benefits of those products to internal teams and stakeholders, is critical to innovation success.

Product innovation is not a one-person job, nor does it happen in isolation within a single team or organization. It requires collaboration across multiple departments and stakeholders within an enterprise. The good news is that with Commerce.AI's **Slack** integration, product teams can easily interact with team members remotely, making innovation much more streamlined.

The goal is to understand whether the customer needs aren't being met, identify opportunities for disruption, test new ideas with potential users early in the process, define and execute on a go-to-market strategy, and then scale once you've validated your assumptions about market needs and desire for your product or service.

To do any of this effectively – especially when it comes to product innovation – you need a way to easily communicate across departments while still maintaining control over who has access to certain information at any given time. This is where tools such as Commerce. AI come in handy; they enable cross-departmental communication that allows teams to collaborate more efficiently.

You should prioritize what information gets shared when the key is to ensure that everyone involved has access to the right information at the right time. This enables every team member engaged in innovation activities within an organization – from engineers building products, through designers creating user experiences, all the way through to leadership deciding which ideas are worth pursuing further – to fully participate in driving innovation forward.

Summary

In this chapter, we learned about the key challenges of consumer electronics brands and how to overcome them by analyzing product data and using Commerce.AI's data engine. These challenges include greater complexity and competition in the consumer electronics space. By analyzing a treasure trove of consumer and product data, consumer electronics brands can uncover insights to overcome these challenges.

AI-based innovation has been disrupting many industries over the last few years. Ultimately, the leading consumer electronics brands are adopting AI to stay at the forefront of innovation.

In the next chapter, we'll explore how to apply AI for innovation in the restaurant industry, which, unlike consumer electronics, has often lagged behind technology trends, but similarly stands to benefit tremendously by using data and AI to their advantage.

7
Applying AI for Innovation – Restaurants Deep Dive

The restaurant industry is notoriously difficult. It's a highly competitive space with slim profit margins, and these challenges have only been exacerbated during the COVID-19 pandemic, in which restaurants are subject to more stringent guidelines than ever before.

The average cost of opening a leased restaurant in the US is $275,000, and only about 20% of new restaurants survive past their first five years. This means that there are many failures, but also many opportunities for those willing to take risks and try new things.

In this chapter, we'll discuss how **artificial intelligence** (**AI**) can help overcome some of these challenges. We'll look at how AI can be used to improve operations and increase revenue, all while reducing costs and increasing profits. We'll also discuss how AI can be used to create more engaging experiences for guests in the restaurant industry.

In this chapter, we are going to cover the following main topics:

- Understanding the challenges of restaurants
- Analyzing product data for restaurants
- Using **Commerce.AI** for restaurants

Understanding the challenges of restaurants

The restaurant industry is one of the most staid and stagnant in retail, with little appetite for change. Think about it: historically, restaurants have had a tough time competing with other forms of entertainment such as TV and movies, or even just socializing with friends. Why would you go out when you could stay at home, order in from **Seamless** or **Just Eat**, and watch Netflix?

Traditional brick-and-mortar retail is struggling mightily to reinvent itself as consumers shift their spending habits online. Meanwhile, the restaurant industry has remained largely unchanged since its inception – there's not much innovation that can occur once a business model becomes so ingrained in society.

But times are changing. In recent years, data has become democratized – accessible to anyone with an internet connection and a computer – allowing companies to make better decisions than ever before based on evidence rather than gut feel or guesswork.

To better understand how and why to use data and AI in the restaurant industry, let's explore some of the major challenges in this industry today.

Profitability

The first major challenge is generating enough revenue to break even. While the end goal is to become (or to stay) highly profitable, this starts with matching the break-even point. While profitability is a universal challenge for businesses, this is particularly true for the restaurant industry, which faces notoriously slim profit margins.

It is estimated that the average profit margin of a restaurant is approximately 3% to 5%, with many restaurants struggling to break even. This has led some to call the restaurant industry a *low-profit* business. So, what does it take to generate enough revenue to break even? And more importantly, how can you increase your revenues in order to achieve profitability?

We have broken down this topic into four key areas: **customer acquisition costs** (**CACs**), menu innovation, pricing strategy, and marketing tactics. Each of these areas will be covered in detail in the following subsections.

Customer acquisition costs

This first area of focus deals with all of the costs associated with acquiring new customers – or CACs for short. These include advertising costs (both online and offline), as well as any discounts or incentives you may offer existing customers in order to bring them back into your business on a recurring basis.

In other words, if you are relying solely on word-of-mouth referrals to acquire new customers, then you will need to invest heavily in marketing efforts so that people know about your location and offerings.

It is also important here not only to consider your customer acquisition costs but also those of your competitors; if there are heavy discounts being offered by other businesses at nearby locations, then it may be wise for you to match or beat those offers.

You should also be aware of any local or state incentives that may be available to help offset the costs of your restaurant. For example, if you are in the process of opening a new location, then there may be grants and tax breaks available that can reduce your CACs.

However, all of this is just a starting point, and using data from the restaurant industry as a whole can help you to refine your marketing and acquisition strategy. In other words, industry-wide CAC data can be used to improve your own CACs, and ultimately your profitability.

Menu innovation and product development

The next major factor impacting profitability when running a restaurant is actually changing what you serve in terms of food. If you are currently serving the same old dishes night after night while hoping for an uptick in sales, then the chances are things aren't going according to plan.

As such, making sure that your menu remains fresh and exciting from month to month is paramount here; offering items beyond basic American fare can help ensure that people continue returning again and again.

Menu innovation is the process of developing new and exciting food and beverage offerings for existing or new customers. It's an essential tool in the restaurant industry, enabling restaurants to stay relevant in a crowded marketplace by offering something different from their competitors – something that will make them stand out from the crowd.

In order to understand how to improve your restaurant menu, we need to step back and look at how this concept of the *customer experience* has taken over our lives. Many people these days are living their lives online, on social media platforms such as **Facebook** and **Instagram**. They're also sharing their thoughts about restaurant dishes and menus, and tapping into this data can provide valuable insights into menu engineering, as we'll explore later in this chapter.

Pricing strategy

The third area of focus when it comes to increasing profitability pertains to the pricing structure of your business – and more specifically, the *why* behind what you are charging for your products or services. If you are currently charging too little for your offerings, then the chances are that this is resulting in lost revenue and a decreased overall bottom line.

To put it simply, if people aren't willing to pay enough for what you have to offer, then they probably won't be returning for future purchases – no matter how good those previous purchases were. You can increase profits by raising prices on selected items that are in high demand while simultaneously lowering prices on other items that may not be as highly sought after. This will help ensure that the balance between the sales volume and average ticket price remains favorable over time.

Be sure to also take into consideration any promotions or discounts being offered by competitors (or possibly even by yourself in the past) as well as any tax implications associated with any such changes. There may be opportunities here to further increase profitability based on tax laws or certain government incentives currently in place. Most importantly, however, is analyzing pricing and demand data at scale to calculate the optimal pricing strategy.

Marketing tactics and social media

Leveraging social media effectively is an increasingly important aspect of marketing efforts today – particularly within the restaurant industry itself. Facebook has become a powerhouse when it comes to reaching potential customers, with over 2 billion users worldwide.

Organic traffic refers to traffic that a company would have gotten absent of paid advertising. Paid ads can actually drive organic traffic as a secondary effect by leading to word-of-mouth referrals. Using **Facebook ads** can help drive foot traffic into your restaurant through organic means, which will ultimately result in higher customer counts – especially when you consider that nearly one-third of visitors who come into brick-and-mortar establishments do so via word-of-mouth referrals from friends or family members first.

By leveraging Facebook ads along with other paid social media campaigns (for example, on **Twitter** or **Instagram**), it is possible to reach a vast audience while driving significant increases in sales over time. Just keep in mind that spending money on paid social media campaigns alone may not necessarily result in increased revenues, as there are many other factors to consider.

Changing guest preferences

Understanding and anticipating consumer behavior is perhaps the most important aspect of maintaining a competitive edge in today's dynamic restaurant marketplace.

The hospitality industry has been through many changes over the past couple of years, and while innovations such as augmented reality have brought new excitement and appeal to dining out, we believe that consumers are looking for more than just a new gimmick. In order to create long-term value for brands, they need to understand what guests expect from their experience – and how those expectations are changing.

Restaurant customer preferences have changed dramatically. Just a couple of years ago, the focus was on aspects such as the taste of the meal, the variety of the menu, the wait time, and the restaurant ambiance. Today, rigorous health and sanitation measures are foremost on customers' minds, and many customers will complain if standards are slack.

These changing guest preferences impact profitability and forecasts as well. For example, there's a new reality of smaller dine-in sales overall, as well as decreased check sizes. As a result, restaurants have to look for new ways to bring in revenue, in addition to satisfying customers based on their new preferences. For example, restaurants can consider changing their menu or even implementing technology to better help them adhere to social distancing protocols (for example, contactless payment options and QR codes for menus).

Preferences come down to context. What do guests see? Who are they with? How hungry are they? Where are they dining? And what else are they doing during their visit (for example, shopping at nearby stores)? These contextual factors play an important role in determining whether an experience will be memorable or not, and whether guests will want to share it with others or not (hence why engagement metrics on social media become so important).

The key takeaway here is that every guest has different needs based on their individual situation, which means that no two experiences will be exactly alike. To stand out amidst today's crowded landscape requires both creativity and innovation: you need something different that your competitors don't have – access to massive amounts of product and service data.

Creating profitable menus (and pricing)

The restaurant industry has been through a lot. Pre-COVID-19, the industry saw decent growth, which plummeted dramatically in 2020, putting many small restaurants out of business. The response of federal governments flipped the switch, driving an increase in discretionary income among consumers and a desire for people to eat out more frequently post-COVID-19. As a result, many new restaurants were opened across the country, and there was an explosion of innovation in terms of technology, design, and food quality.

In order to find success in this tumultuous environment, restaurants must adapt and evolve in order to stay relevant to today's consumers.

Menu engineering

Menu design is crucial to the success of a restaurant. A well-designed menu will increase customer loyalty, build brand identity, and enhance overall revenue growth. Unfortunately, many restaurants struggle with menu design. There are a number of reasons for this, but one of the most common is that menus often contain outdated information.

Restaurants relying on legacy technology – whether it's paper menus, tablets, or mobile apps – won't be able to keep up with today's pace of change in the restaurant industry. In order for restaurants today to compete effectively in an increasingly competitive landscape, they need new ways of communicating with their guests and creating engaging experiences around what they eat.

Taking it to the next level, restaurants don't only need to design menus, they need to *engineer* menus that are loved by consumers while maintaining high profitability. Menu engineering is one of the most complex and challenging strategies for creating profitable restaurant menus. To do it right, restaurants must be innovative in both product development and marketing.

Innovation in product development is critical because it's the only way to keep up with changing consumer demands, while still providing variety and ensuring customers are satisfied. As consumers become more health-conscious, they're looking for more natural ingredients in their food – and they'll pay a premium for these products if you include them on your menu. This means that you need to invest in developing new products that have a real chance of becoming best-sellers.

Maintaining online reviews and social media marketing

Online reviews and social media have become an essential part of the consumer's decision-making process when choosing where to dine out. However, with the growth of online review platforms and social media channels, restaurants are finding it harder to maintain their online presence. Online reviews and social media posts are no longer just a matter of informing consumers about a business; they have become a way for consumers to interact with brands on their own terms.

Online reviews and social media posts can be powerful tools that help differentiate a restaurant from its competitors – or can backfire if not handled properly. For example, negative reviews or bad press can damage a brand's reputation and could lead to customers avoiding the company in the future.

The advent of online review platforms has made it easier than ever for people to voice their opinions about businesses, but maintaining an online presence is no easy task – especially when competition is high and there are so many other options for dining out.

Now that we understand some of the major challenges facing restaurants, including profitability, maintaining online reviews, and social media marketing, let's explore how restaurants can analyze product data to overcome some of these challenges.

Analyzing product data for restaurants

The restaurant industry is a highly fragmented market with an enormous variety of products and services available to consumers. As a result, it can be difficult for restaurateurs to understand the data they need to drive meaningful innovation in their businesses.

This section introduces several different ways that data can be used as a tool for innovation within the restaurant industry, including predicting food item success, predicting competitor performance, new profile discovery, and more.

Predicting how food items are likely to perform

One big way to use data for restaurant innovation is to predict how food items will perform in-market. By analyzing the performance of food items in the restaurant industry at large, restaurateurs can identify which food items are likely to be most popular with consumers. This information can then be used to inform business decisions about menu items, pricing, and marketing campaigns.

Traditionally, restaurants would rely on their experience to make business decisions. However, with more data to hand, restaurateurs can use data science to gain insights into how customers are likely to respond to food items. For instance, a restaurant might use Commerce.AI to automatically monitor social media sentiment surrounding its dishes and adjust its menu accordingly.

In the last few years, we've seen increased investment in data science and AI across a wide variety of industries. However, in the restaurant industry, the use of these tools has been relatively slow to develop. The main reasons for this are that it is difficult to predict how individual consumers will respond to new trends or innovations in real-time and that it is costly and inefficient for restaurants to collect large amounts of consumer data.

However, with rapidly advancing technology, we now have the ability to overcome both of these challenges. And as a result, more and more restaurants are starting to leverage data science and AI in their operations.

Predicting how competitors will perform

Another way that data can be used is by taking a look at historical sales information for similar products or service offerings from competitors. This information can then be analyzed to determine the most promising market niche for a new food or beverage offering, or a new product or service more broadly.

For example, if one of a restaurant's competitors has recently launched a new dessert item that seems to be selling well, the restaurant may consider introducing a similar dessert item as well. This approach allows restaurateurs to learn from their competitors and take advantage of what they perceive as untapped opportunities within their industry space.

Competitive intelligence is an important tool for any business, but it can be especially valuable for those in the food and beverage industry. The competitive landscape within this space is constantly changing as new competitors enter the market, and existing ones adapt their offerings to remain competitive.

In order to stay on top of these changes, restaurateurs need to have a strong understanding of what their competitors are doing so that they can make informed decisions about how best to position themselves in the market.

Predicting customer needs based on previous purchases

Further, data can be used to help make informed predictions about what customers want based on their previous purchases. For example, by looking at where people buy different types of wine or beer, it's possible to determine which types of wine or beer are likely to be most popular within certain markets/regions/sectors (for example, reds tend to sell better than whites in the US).

Using this information along with other consumer data, such as age group and purchasing patterns, can help restaurants develop more targeted marketing campaigns that appeal directly to customers who have shown a propensity toward certain types of products or brands before.

Data collected from previous purchases can also provide insights into what motivates certain **customer segments** within each **market segment** (for example, millennials prefer beer over wine). By understanding how different customer segments react differently depending on the context of their purchase decision (such as whether they're buying for themselves or as part of a gift basket), it's possible to create more tailored promotions.

New profile discovery

Many of the products and services available in the restaurant space are relatively new. For example, many consumers have never eaten a sweet yogurt bowl before, or they've never tried a high-protein food that is baked instead of fried.

In order to create meaningful innovation, restaurateurs need to be able to understand their customers' behavior and preferences across a variety of dimensions (such as food type, experience type, and price point) in order to find opportunities for improvement and growth.

Data can uncover new insights about the profile of consumers who visit a restaurant. For example, data analysis may find that sweet-toothed guests are more likely to spend more money on average than those with a savory appetite. This insight could be used by restaurateurs to understand which dishes or products within their menu are performing well and why, potentially leading to new product development or service offerings.

Ultimately, data can be a powerful tool for restaurant innovation. From predicting food item success and competitor performance to new profile discovery, data can enable meaningful competitive advantages. Next, let's look specifically at how restaurants can use Commerce.AI for boosting innovation.

Using Commerce.AI for restaurants

In previous chapters, we've looked at how to use Commerce.AI for various product-oriented industries, including luxury brands in *Chapter 4, Applying AI for Innovation – Luxury Goods Deep Dive*, wireless networking brands in *Chapter 5, Applying AI for Innovation – Wireless Networking Deep Dive*, and consumer electronics brands in *Chapter 6, Applying AI for Innovation – Consumer Electronics Deep Dive*.

As we'll see, Commerce.AI can be used in the service industry as well, including for restaurants. This is because the service industry is highly data-rich, as customers leave large amounts of both implicit and explicit feedback.

In fact, so much data is generated, largely in the form of customer feedback, that it would be impossible to manually sort through and analyze a meaningful amount of it. That's where AI comes in, which automatically finds trends in customer reviews and can make predictions about the future.

Let's look at five main ways Commerce.AI can be used in the restaurant industry:

- Analyzing restaurant customer data
- Mobile surveys
- Gauging customer sentiment response based on marketing campaigns
- Staying connected with customers
- Restaurant trend analysis

Finally, we'll look at a case study of how a French pizza chain used Commerce.AI for restaurant innovation.

Analyzing restaurant customer data

As one of the leading players in the restaurant technology space, with over 1 trillion data points processed through its platform (from tens of thousands of products and services), we've seen firsthand how companies leverage our platform to get insights into their restaurant businesses. This experience has helped us develop some best practices on how restaurants can effectively leverage our services when doing business analytics projects.

Analyzing restaurant customer data is no different, and most restaurants can benefit from taking a deep dive into their data to understand their customers. It is also useful for understanding how your food and beverage consumers create value that you can leverage in your business – whether it's managing operations, developing new products, improving marketing campaigns, or all of the above. But before we dive into how to do it, let's take a look at some of the challenges that most restaurant businesses face when doing analytics in their businesses.

When you think about analytics in your restaurant business, you might begin with company-wide reporting – think revenue by unit or revenue by shift – or product improvement such as finding the optimal inventory levels (including position) for each item on a menu board. Both are great uses of an analyst or manager's time because they're actionable insights that impact decision-making immediately.

Unless the goal is to find further correlations between elements within your existing dataset(s), these kinds of reports won't provide much insight into repositioning products or changing spending behaviors.

That's because they use pre-existing patterns without challenging those patterns in any way; therefore, there's very little beyond correlation/causation found, and nothing gained from putting things into more meaningful context through analysis (such as combining all shift information). It feels like just throwing more irrelevant detail onto an already complex pile that nobody understands but everybody has to deal with.

Commerce.AI solves these issues by combining internal, existing data with large amounts of external market data and summarizing findings in quick and snappy insights.

Mobile surveys

Mobile-based surveys have become an essential tool for understanding consumer behavior in relation to your restaurant products and services.

Mobile surveys are a great way to understand how consumers use your restaurant, as well as what they like and don't like about it. They can also be used to understand how consumers feel about the quality of your food, whether you're keeping them happy with their experience, and if you need to make any changes.

In the restaurant business, the customer is king, and with mobile surveys, you can have an intimate look at what your customers want.

Mobile surveys can be completed quickly and easily in Commerce.AI. They're a great way to understand your consumers in real time, as well as to set up regular *health checks* on the quality of your food or service.

The data from these surveys can also be used to inform more traditional forms of marketing such as print ads, social media posts, and organic **search engine optimization (SEO)**. You could even create a custom survey for specific audiences via email and put it in front of them when they come into your restaurant.

Gauging customer sentiment response based on marketing campaigns

Marketers spend a lot of time and energy on developing and executing initiatives to increase brand engagement with our customers. The question they often ask themselves is *does this effort pay off?*

We can measure engagement, sales lift, and other metrics to determine if the investment was worth it, but there are many factors that influence customer behavior that make it challenging to accurately predict how a campaign will perform. One data source that can be immensely helpful in understanding how an initiative will play out is sentiment analysis.

Sentiment analysis uses **machine learning** algorithms to identify whether the written content is positive, negative, or neutral in nature. Since social media posts are generally longer than traditional marketing campaigns, they give us more information about what types of things our customers like or dislike about our products and services.

This allows us to quickly understand the overall sentiment of a campaign and make adjustments as necessary before investing too much more money in an initiative.

Commerce.AI provides powerful tools for sentiment analysis because it integrates with popular third-party services such as **Google Analytics** and **Facebook Insights**, two widely used tools when measuring customer engagement with online businesses. These integration capabilities let you analyze customer sentiments using the same data sources that you use for your marketing analytics programs, which means you can begin making smarter decisions about how to build your business from day one, instead of building it after the fact based on feedback from existing customers.

Stay connected with your customers

Social media has become an important channel for staying in touch with customers. Companies are engaging with their followers on Twitter, Facebook, Instagram, and other platforms in real time so they can learn what matters most to them as consumers. This knowledge helps brands make smarter decisions about their products and services, grow sales, improve customer loyalty programs through personalized offers, and increase engagement across all channels.

Monitoring social media conversations around a given brand or competitor is a great way to stay connected with your customers. This means that you can hear about any emerging issues before anyone else does and respond quickly when something goes wrong or needs fixing. It's also a great way to get alerts when potential new customers show interest in your brand or product category.

You've probably heard the phrase social media **return on investment** (**ROI**). It means measuring the benefits of social media campaigns, whether it be through likes, comments, shares, or other interactions on your company's social platforms. With social media as a platform for customer engagement, brands can listen in on what their customers are saying and learn from their feedback.

If you have a customer service team that is consistently answering questions from customers using social channels such as **Facebook Live** or **YouTube Chats**, you should be able to track how many people watch the live stream and interact with the content. A similar approach can be taken when listening to comments that customers leave on your website or blog posts. By tracking comment data, you can see who is engaging with your brand and what they are saying about your company and products.

The key takeaway here is that companies can use real-time data analysis tools such as Commerce.AI to stay connected with customers across multiple channels so they can understand what types of content resonate most with them and why. This allows companies to create targeted content marketing campaigns that keep their audiences engaged and ultimately drive more revenue for their businesses.

Finding and predicting trends in the restaurant business

As part of your innovation strategy, you should be constantly looking for new ways to add value to the business. The restaurant industry offers many opportunities for innovation and growth – but you need a way to find those opportunities so that your team can execute on them.

Innovation teams can use Commerce.AI to identify potential trends in the restaurant industry and potential innovations that can add value for your customers.

The goal of any good analysis is not just to spot interesting trends, but also to make *predictions* about what will happen next based on those trends. For example, if there's been a recent increase in the number of reservations made for Sunday brunches, then it might be a good time for an innovation team within your restaurant chain to release a new menu item specifically targeted at these types of high-value bookings.

If you can predict what people want before they know they want it themselves, then you have a competitive advantage. This is why innovation teams need to be data-driven. They need to know what their customers want so that they can build and market products or services that are designed to meet those customer needs.

In order for innovation teams to create the products and services that will bring customers flocking, they need a thorough understanding of how people behave when they eat out. This requires them to look at trends in the restaurant industry, which has been experiencing an unprecedented amount of change over the past decade.

A case study – how a large French pizza chain used Commerce.AI

A leading French pizza chain used Commerce.AI to analyze customer feedback at scale and evaluate both store service and product quality. This French pizza chain evaluated over 100,000 customer reviews across 385 stores to analyze metrics including the following:

- Overall review sentiment
- Store leaderboard – best and worst stores
- Top attributes
- Comparing stores across attributes

Let's dive into each of these areas.

Overall review sentiment

Restaurant reviews are a huge part of the online journey for consumers when considering where to eat. In fact, consumers spend a significant portion of their time researching restaurants before deciding where to eat.

But this also means that businesses need access to reliable data on how their customers are engaging with their content. And nowhere is this more critical than in the world of online reviews, where an average consumer checks out several reviews before making a decision about which restaurant or experience to engage with next.

The good news? There's now an open standard for measuring the sentiment on all sorts of review platforms – and it's called *using Commerce.AI*. Traditionally, businesses only had two options when it came to using customer feedback: ignore negative sentiments or respond quickly and personally (often using bots).

This has resulted in a lot of conflict on review platforms – negative sentiments coupled with personal replies from business owners who feel compelled to defend themselves against unfair accusations. In essence, businesses have been fighting fire with fire rather than using data insights from customer feedback as a tool for improving their products and services.

With Commerce.AI, brands can quickly and easily measure the overall review sentiment. The French pizza chain was able to gain immediate insights into the overall review sentiment, informing them how their customers were feeling at any given time.

Store leaderboard

Large restaurant chains often have hundreds of stores and millions of customers, making it difficult to track performance at an individual level. By applying sentiment analysis and AI to data collected from social media platforms such as Twitter, the pizza chain innovation team was able to identify the top-performing stores by measuring how happy customers were with their service and food quality, as seen in *Figure 7.1*:

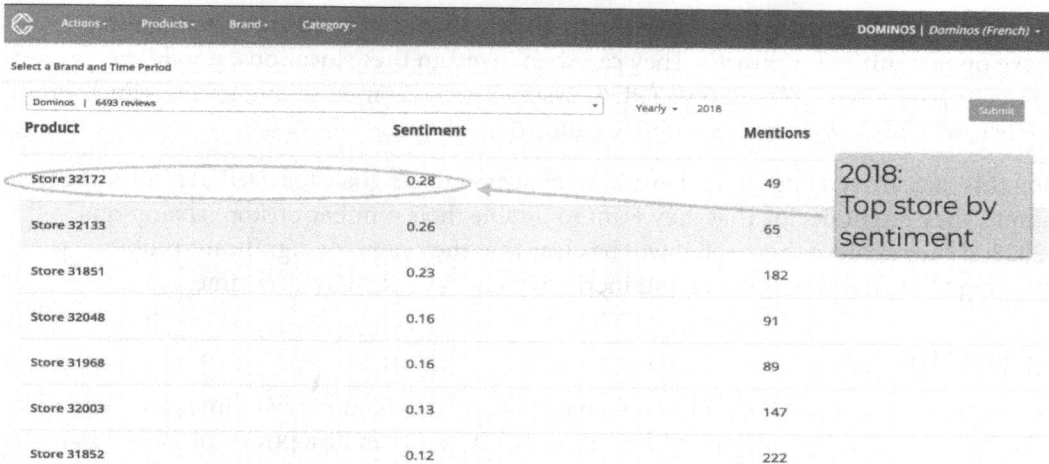

Figure 7.1 – Leaderboard of top stores by sentiment in Commerce.AI

Similarly, this data and AI analysis indicate the worst-performing stores, as seen in *Figure 7.2*, allowing them to quickly address areas for improvement. The goal is to learn from the stores leading the way and apply those lessons across the board while pulling up the stores that are faltering, improving brand image and customer satisfaction from both sides. After all, each and every store plays an important role in your overall brand.

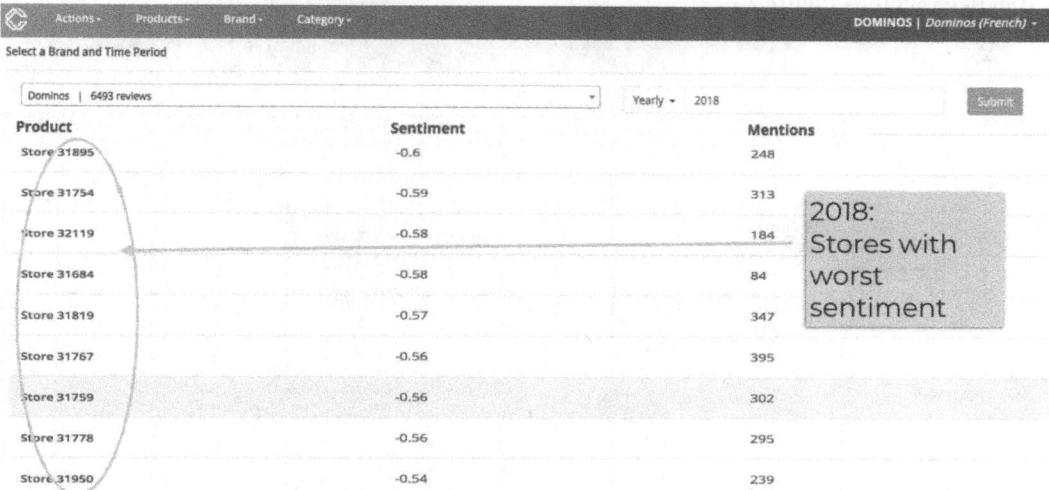

Figure 7.2 – Leaderboard of worst stores by sentiment in Commerce.AI

By using this approach, the team is able to identify which stores are underperforming or have opportunities for growth. They can then invest in these locations, providing resources such as new **Point of Sale** (**POS**) systems and training for employees, which can help turn around performance at these locations.

Using this data-driven approach also allows the chain to set goals for itself over time. For example, they might decide that they want to double their number of store champions by 2025; if they achieve this goal, it will be clear that they've made significant progress in improving their network of stores and increasing customer loyalty over time.

Top attributes

Understanding the specific attributes of a particular location and how to improve them is the Holy Grail for many companies. If used correctly, data can be a powerful tool to gain an edge in competition.

Using AI to measure store attributes such as price points, product quality, or customer service can give retailers and restaurants an advantage over their competitors.

However, this requires gathering large amounts of data from multiple channels. The pizza chain used Commerce.AI to analyze data across hundreds of its locations and tens of thousands of customer reviews to measure and analyze specific attributes for each individual store, highlighting areas for improvement for a ground-up innovation approach, as seen in *Figure 7.3*:

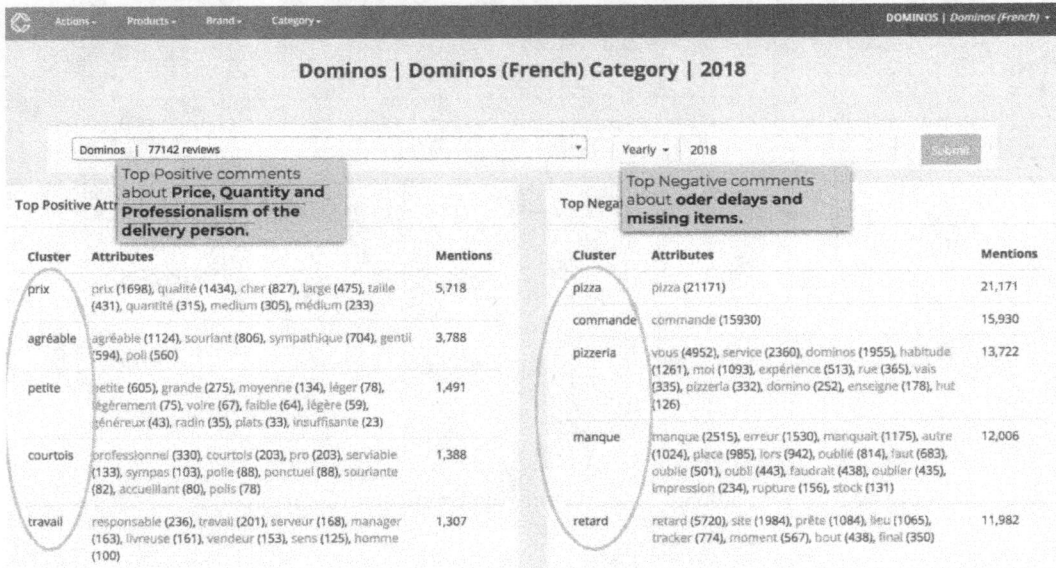

Figure 7.3 – Analyzing store attributes in Commerce.AI

In this case study, most attributes are in French, as we're analyzing data from France. We can see that positive attributes reference things such as price, quantity, and the professionalism of the delivery person, while top negative attributes reference things such as order delays and missing items.

Comparing stores across attributes

Each of the 385 stores analyzed by the pizza chain had a unique attribute profile, with varying sentiment across attributes such as price, quality, professionalism, and cleanliness. Using the Commerce.AI data engine, we were able to compare each of these attributes between stores in order to find specific areas for improvement for any given store, as seen in *Figure 7.4*:

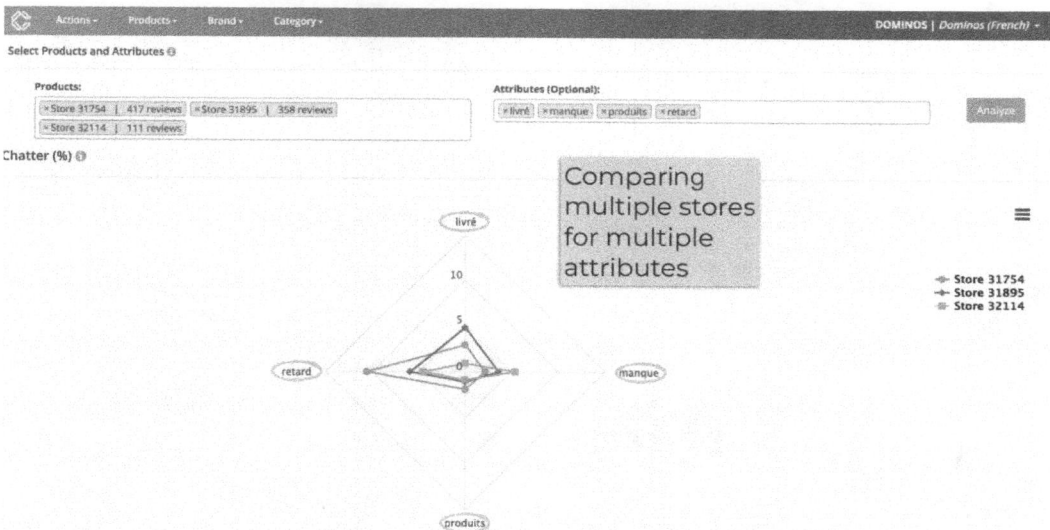

Figure 7.4 – Comparing multiple stores for multiple attributes

We can see, for instance, that `Store 31754` has particularly intense chatter for the French word `retard`, or *delay*, indicating that measures need to be taken to make processes more efficient at that store.

With the power of data and AI, the pizza chain was able to gain unparalleled insights into the sentiment of their stores and competitors' stores, as well as discover opportunities to meet consumer wants and needs.

At a high level, these insights can be consolidated in the form of data-driven reports, which provide a quick overview of the data for innovation teams to understand the hearts and minds of their consumers. In *Figure 7.5*, for instance, we can see a mockup of what this dashboard may look like for a pizza chain, the data of which would be populated by the specific brand and stores selected.

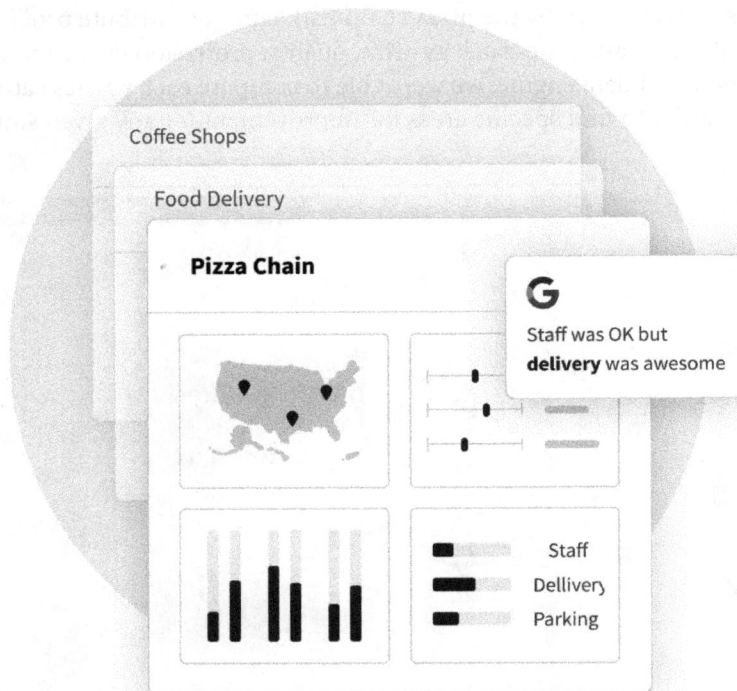

Fig 7.5 – A Commerce.AI mockup of a high-level consumer insights dashboard

We can see quick summaries of the data from thousands of reviews, such as **Staff was OK but delivery was awesome**, highlighting an opportunity to improve teams and market their excellent delivery processes.

In conclusion, Commerce.AI is a versatile tool that restaurants can use to analyze customer data, deploy surveys, gauge customer sentiment, analyze trends, and more. By implementing these kinds of AI use cases, restaurants can get ahead of the competition that's stuck using traditional analytical methods (or even worse, little-to-no analysis at all).

Summary

In this chapter, we've explored how data is a critical tool for restaurant innovation teams. Using data to inform your restaurant strategy is an essential part of being successful in this evolving industry. The right data, at the right time, can save you countless hours of trial and error, help you make smarter decisions, and even help determine what food and beverages to sell or services to offer in the first place.

AI isn't just about automating repetitive tasks; it's also about providing insights that no human could have come up with on their own. These tools are perfect for the unique challenges faced by restaurant owners, from managing inventory across multiple locations to understanding consumer behavior on social media platforms such as Instagram.

As consumers engage with food in new ways through digital channels – apps such as Instagram or **Pinterest** – restaurants have an opportunity to reach them directly with offers or content that they find interesting or engaging enough to share with their friends on these channels.

In the next chapter, we'll go through one more industry deep-dive, and look at how consumer goods businesses can apply data and AI for innovation. Like restaurants, consumer goods businesses are facing stiff competition, and they can use data and AI to overcome key challenges.

8
Applying AI for Innovation – Consumer Goods Deep Dive

The consumer goods industry is an important part of the global economy. It's worth more than $635 billion in the US alone – a figure that is only going to increase as consumers become wealthier and have greater access to products and information.

Since the global financial crisis of 2008 and the more recent **COVID-19** pandemic, the consumer goods sector has faced a number of headwinds. The big three – shopping, travel, and leisure – are all under pressure. Fewer people are going out to eat or visiting malls, and more are choosing to stay in and watch TV instead.

At the same time, e-commerce is booming and even quickly overtook in-store sales due to the pandemic. And with consumers becoming increasingly comfortable with digital platforms for shopping, the growth of e-commerce is set to continue unabated.

As such, consumer goods companies that fail to adapt risk losing customers altogether, as shoppers seek out new ways to buy their favorite products quickly and easily online.

In this chapter, we'll explore how consumer goods firms can use data and **artificial intelligence** (**AI**) to innovate and adapt to these changing trends, enabling them to stay ahead of the competition.

In this chapter, we'll cover the following topics:

- Understanding the challenges facing consumer goods brands
- Analyzing product data for consumer goods brands
- Using **Commerce.AI** for consumer goods brands

Technical requirements

You can download the latest code samples for this chapter from the book's GitHub repository:

```
https://github.com/PacktPublishing/AI-Powered-Commerce/tree/
main/Chapter08
```

Understanding the challenges facing consumer goods brands

The consumer goods industry has been under significant pressure in recent years from rising input costs, changing consumer habits, and new technology, as well as other factors that are outside of its control (for example, supply chain uncertainty).

Companies within the industry must operate in an increasingly complex regulatory landscape while also dealing with increasing competition from online players, including e-commerce platforms such as **Amazon** and **Flipkart**. In today's rapidly evolving world of retailing, no retailer can afford to be complacent or risk falling behind in their market.

As a result, many smaller brands have filed for bankruptcy protection or have seen their stock prices hit rock bottom due to the challenges they face across multiple fronts, including the following:

- Competitive consumer goods
- Consumer goods market intelligence
- Inventory management

- Creating the right product mix
- Creating consumer goods content at scale
- Consumer goods review analysis

Let's dive into each of these challenges in detail, which will help inform the need to use data and AI to stay solvent and ultimately become successful.

Competitive consumer goods

Companies in this industry are competing against each other for market share within a very crowded space, with hundreds of new products launched every year. These companies are also facing an increasingly competitive landscape, with more players entering the consumer goods space and consumers becoming more demanding and discerning about what they buy.

One of the major challenges faced by companies in this industry is that they have to constantly innovate to stand out from their competitors and deliver better experiences for their customers. Innovation is critical for these companies to stay relevant in an ever-changing environment and ensure that they continue to be able to attract and retain customers.

Innovation is a broad term that can mean different things to different people. For some, it might simply mean creating a new product or service, while others may think of innovation as process improvement or organizational transformation. Regardless of the definitions used, all successful consumer goods companies are constantly looking for ways to improve their products to stay ahead of the competition and satisfy customer needs.

Consumer goods market intelligence

Market intelligence is a critical component of successful product innovation in the consumer goods industry. Understanding how to create and leverage insights is essential for creating products that meet consumer needs, and it's also an important tool for guiding long-term strategic decisions about the future of your business.

But what exactly is market intelligence, and how can you use it to help your organization create customer-centric products with the potential to drive growth? Market intelligence is a broad term that describes a variety of different sources of information that can be used to understand your customers.

It includes data from traditional market research methods, such as focus groups and surveys, as well as more creative approaches such as social media monitoring. One important aspect of any market intelligence program is ensuring that it's regularly refreshed – otherwise, you risk falling behind the pace of industry changes and developments.

Understanding your current and potential customers is critical for creating customer-centric products that meet their needs. And while consumer goods companies have long relied on data to inform their product development decisions, many still struggle with the challenge of conducting effective market intelligence programs.

Inventory management

Managing stock is one of the most pressing issues facing consumer goods companies today. With e-commerce, consumers can order products whenever and wherever they want. This has led to a boom in online retail sales, but it also poses a challenge for traditional brick-and-mortar stores that have to compete with on-demand shopping options.

Inventory management has become a key differentiating factor for retail stores. A company's ability to manage its inventory effectively can determine whether it can succeed on its own, become an acquisition target, or fail. For this reason, many companies today are dedicating significant resources to making sure they have the best possible inventory management systems in place.

The consumer goods industry is no different, and there are several factors at play that make managing inventories even more challenging. These include the following:

- Product obsolescence
- Rapidly changing consumer preferences
- Short product life cycles

Let's briefly explore how each of these creates challenges for managing inventory:

- **Product obsolescence**: This one is obvious, but it's worth noting nonetheless. The products you buy now may be obsolete next year – or even sooner if innovations come along and render your product obsolete in just one season. If products become obsolete sooner than expected, say, because of a shift in consumer interest, then you'll be left with excess product.

- **Rapidly changing consumer preferences**: People have short attention spans nowadays and prefer instant gratification over delayed gratification. As such, what customers want today may not be what they want tomorrow – especially as innovations emerge that appeal to them on an emotional level.

- **Short product life cycles**: In many cases these days, consumers see a product for only a few months before they move on to the next thing – which means retailers need to constantly refresh their offerings or risk losing sales. This makes inventory management all the more crucial.

Creating the right product mix

The **product mix** is a hot topic in the consumer goods industry. A few years ago, analysts and investors were obsessed with finding ways to increase product mix by acquiring complementary brands or entering new markets. Now, as we enter an era of changing consumption habits of emerging consumers, the focus on product mix has shifted from simply acquiring complementary brands to creating new ones.

As companies look for new ways to create value in their businesses, one way they're considering is by looking at their portfolio of brands through a lens of product innovation. The idea here is that if they can create more unique products that are differentiated from competitors' offerings – whether it be through attributes such as design or function – then they can grow their businesses by attracting more customers.

This approach also offers some practical benefits; by creating more differentiated offerings across multiple brands, companies can leverage scale and distribution capabilities across all their brands to generate higher returns on investment for shareholders by driving sales growth or increasing margins per sale. This makes sense since you're not only growing your business but creating financial value as well.

It stands to reason that if a company creates more unique products that are differentiated from its competitors' offerings, then it will attract more customers. This is why we believe this shift in thinking around product innovation may lead to an increase in investing in consumer goods companies as well, over time.

Creating consumer goods content at scale

Product pages are an important place for any brand to reach customers and tell its story. This is where consumers can get a better sense of what the brand stands for and how they should use the product.

To drive conversions and increase sales, it's necessary for brands to invest more time and energy into creating engaging product pages than they did just a few years ago. However, creating high-quality content can be difficult due to time constraints and limited resources.

In the next section, *Analyzing product data for consumer goods brands*, we'll discuss how companies can leverage data science tools such as **machine learning** in order to create effective product pages that will help them win over customers online – ultimately increasing sales volumes across the board.

Consumer goods review analysis

Product reviews are one of the most popular and extensive user-generated data sources for many companies. The majority of consumers' purchase decisions are influenced by product reviews.

Product reviews are also a major source of *competitive intelligence* for consumer goods manufacturers – they can learn what products consumers like best, as well as what features and functions consumers value the most.

Companies use this information to inform their product development and market strategy. For example, if a company is launching a new deodorant that uses plant-based active ingredients, it's likely that they'd look into existing customer reviews about other brands' deodorants for insights on how consumers perceive this new technology.

In order to gain actionable insights from these reviews, companies have traditionally relied on manual methods or paid third parties to manually help with the review process. However, there are now several tools on the market that aim to automate some or all aspects of analyzing product review data online.

Traditionally, automated **sentiment analysis** has focused on identifying negative language (euphemisms used instead of the word *boring*, for example) in product reviews. This approach generates an opinion score that represents how positive or negative a reviewer feels about a given product based on its content.

The problem with this approach is that it misses many subtle nuances in reviewer language, which can lead to skewed results when applied consistently across different industries or review platforms.

Furthermore, it's not always clear whether certain terms used by reviewers are intended as compliments or criticisms (for example, *that's just great*). A more nuanced approach is needed! Reducing bias in automated sentiment analysis requires training datasets with diverse examples drawn from multiple industries and review platforms – but where can we find such a diverse set? Machine learning is the answer, and in the following section, we'll explore how to use it.

Now that we understand some of the major challenges facing consumer goods brands, let's explore how to analyze product data to overcome these challenges and get ahead.

Analyzing product data for consumer goods brands

The elephant in the room when it comes to innovation in consumer goods today is data – there are vast amounts of customer data available today. So why aren't all consumer goods companies using this data to develop better products? It turns out that many don't know where to start with product innovation, given the complexity involved and the lack of available resources within their organizations.

The following are a few of the main ways in which consumer goods brands can use product data:

- Consumer goods content generation
- Analyzing consumer goods reviews
- Lead time analysis
- Demand forecasting
- Maintaining adequate cash flow
- Analyzing the impact of discounts
- Identifying seasonal trends
- Social media analytics

We will explore these methods in-depth in the following sections.

Consumer goods content generation

Creating clear, concise product copy is key to successful products. After all, consumers have millions of products to choose from, but only a limited amount of time and attention. This massive volume of products is a double-edged sword – it means that product teams have to spend time writing copy across multiple marketplaces, and constantly making updates based on new features and product iterations.

All this takes time away from product innovation, but the good news is that it can be automated with AI. In particular, we can create a product copy generator powered by **natural language processing** (**NLP**) algorithms using GPT-J.

GPT-J is a large language model, or a machine learning model that was trained on large amounts of text, released by a group called **Eleuther AI**. We'll demonstrate it as follows:

1. First, install GPT-J and import the required library:

```
!pip install gptj
from GPTJ.Basic_api import SimpleCompletion
```

Since these large language models use pre-training, the model was already trained on a lot of text data, and we only need a small amount of data to tune the model for a specific task.

2. Next, we define this task by providing prompt, which includes examples of product descriptions being generated from a product name and features:

```
prompt = "Write one sentence descriptions for products
based on a list of features.\n##\nProduct: Sundef\
nFeatures:\n- Sunscreen for athletes\n- Unique formula
to prevent burning eyes\n- Can be worn on the body and
on the face\nOne sentence description: Sundef face &
body sunscreen for athletes keeps your skin protected
without hurting your eyes, so you can keep your head in
the game.\n##\nProduct: " + product + "\nFeatures:\n" +
features + "\nOne sentence description:"
```

This prompt is crucial for the language model because language models have a broad range of use cases, including classification, generation, translation, transformation, and more. So, they have to be guided to complete specific tasks. Using the prompt variable, we can guide the model to act as a product copy generator.

3. Further, we'll want to pass a number of parameters, primarily temperature (or randomness), max_length (or the maximum output size of the model), and product (or what the user types in, such as SlimWallet):

```
temperature = 0.4
top_probability = 1.0
max_length = 5
product = "SlimWallet"
```

4. Finally, we can now pass the `prompt` variable and the parameters to the model to create a recommendation. We'll also grab just the first line of text generated, in case the model goes overboard:

```
query = SimpleCompletion(prompt, length=max_length,
    t=temperature, top=top_probability)
Query = query.simple_completion()
lines = Query.splitlines()
results = []
```

In doing so, giving the model an input about a wallet with three features generates product copy such as *SlimWallet is the most stylish*, *thin*, and *durable wallet you've ever seen*.

There are other ways to try out this same concept even without using any code at all, such as with **AI21 Studio** (`https://studio.ai21.com`). In *Figure 8.1*, we use the same prompt in a visual canvas instead of code, and given the `SlimWallet` item, AI21 Studio recommends `SlimWallet holds 15 cards in a thin shape that's 5 times thinner than a traditional leather wallet`.

Canvas ⑦ Quickstart   🗑 Clear all   ↝ Share

Write one sentence descriptions for products based on a list of features.
##
Product: Sundef
Features:
- Sunscreen for athletes
- Unique formula to prevent burning eyes
- Can be worn on the body and on the face
One sentence description: Sundef face & body sunscreen for athletes keeps your skin protected without hurting your eyes, so you can keep your head in the game.
##
Product: SlimWallet
Features:
- Made of top quality, abrasion-proof fabric
- 5 times thinner than a traditional leather wallet
- Holds up to 15 cards
One sentence description: **SlimWallet holds 15 cards in a thin shape that's 5 times thinner than a traditional leather wallet.**

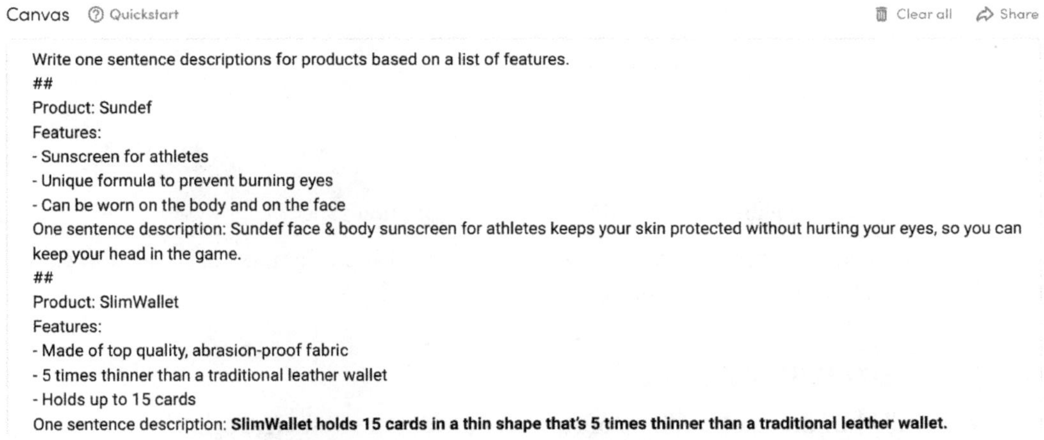

Figure 8.1 – The AI21 Studio canvas for product description generation

As with GPT-J, we'll need to provide a number of settings, which is done in AI21 Studio through a **Configuration** panel shown here:

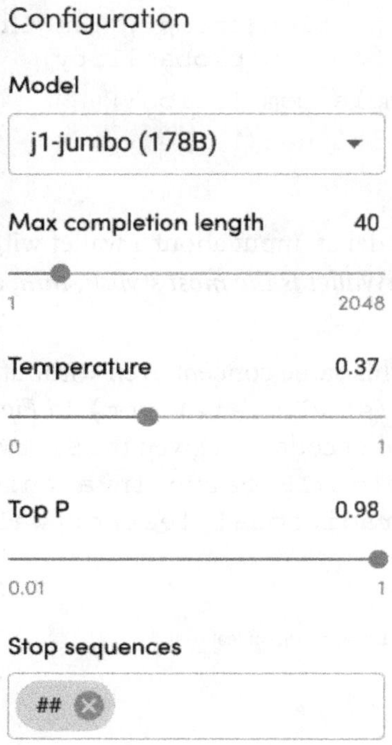

Configuration

Model

j1-jumbo (178B)   ▼

Max completion length   40

1   2048

Temperature   0.37

0   1

Top P   0.98

0.01   1

Stop sequences

✕

Figure 8.2 – The AI21 Studio Configuration panel

The settings (as seen in *Figure 8.2*) are almost identical and include maximum completion length, temperature, and stop sequences.

Analyze consumer goods reviews

There are literally hundreds of millions of Amazon product reviews (`https://nijianmo.github.io/amazon/index.html`), which makes it impossible for any product team to manually read and analyze reviews to extract insights at scale.

Fortunately, we can again use large language models, and automatically extract insights from reviews. Let's look at how to extract user-desired features from a product review. As we've explored in the *Consumer goods content generation* section, we can provide pre-trained language models with a prompt to guide them toward a specific use case.

In *Figure 8.3*, we provide AI21 Studio with a prompt that extracts **Areas for improvement** from a product review:

Review: There is a reason this phone is half the price of other models since it's missing some luxury features. Apparently this phone isn't waterproof, which comes as a surprise to me since I've never known any phone to be waterproof and am always careful with my phones around water. Wireless charging capability is also missing from this phone so you may miss that feature if wireless charging is your thing. It really isn't a big deal though since the battery lasts so long this phone is rarely on the charger anyway.
Areas for improvement: [Waterproof, wireless charging]
###
Review: I just received this and have already experienced video call issues in terms of the quality, on whatsapp and duo. My callers get terribly pixelated videos of me that are so bad that they can't even see my face anymore. And other times it gets so blurry and visually noisy especially in low light settings, with a pixelated areas. I know it is not my internet speed because I've never experienced this before with my last pixel xl phone. I noticed this issue before the Android 11 update but it still continued afterwards. Nothing changes even after I restart. The phone is not even a day old. I'm not sure what to do now. I was so excited to finally receive the phone after waiting for so long, but this is unacceptable.
Areas for improvement: [Video calling]
###
Review: They claim phone battery last 24 hours, not even close. 12hours max so far with data and bluetooth turned off and shuting down some other bells and whistles. I don't use my phone much. I switch off the annoying gestures and put the home button on. It's easier to navigate that way. Spotify has crashed twice so far with this phone.
Areas for improvement: **[Battery life]**

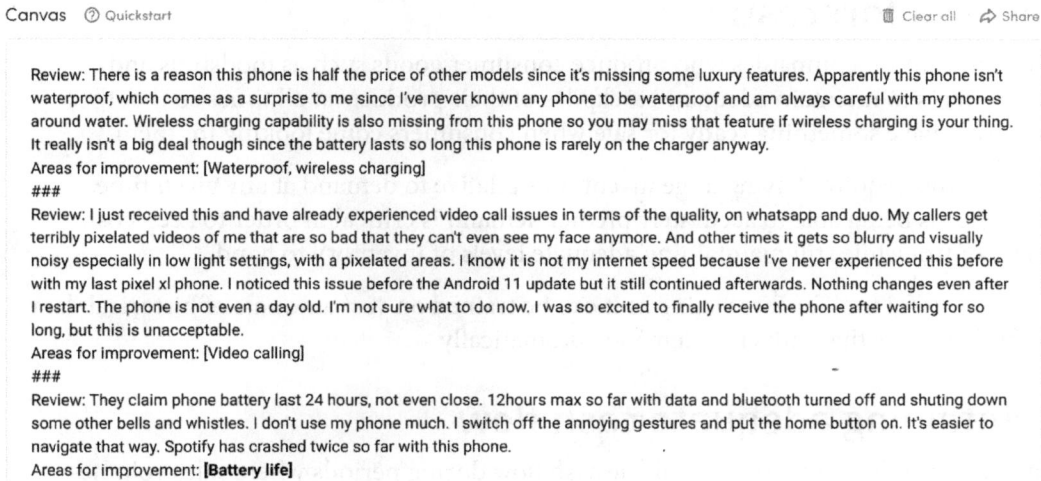

Figure 8.3 – The AI21 Studio canvas for product review analysis

Now, we can provide the model with any product review, and it will extract an area for improvement, such as `Battery life`. We can then quickly scan hundreds or thousands of product reviews and tally up the frequency of any given item in order to prioritize it. For example, perhaps 50 reviews request a better battery life, while only 15 reviews request waterproofing, which would inform product teams to prioritize battery life in the next product iteration.

This process can be scaled to any number of products and reviews, enabling instant insights at scale.

Just as we've generated product descriptions programmatically, we can use this new `prompt` variable (seen in *Figure 8.3*) to programmatically extract areas for improvement from product reviews. An example of this is given in the GitHub repository for this chapter, which can be found in the *Technical requirements* section.

Lead time analysis

Understanding how long it takes for a product to be manufactured or produced can be very revealing about its quality and manufacturing processes. An overly long lead time can indicate that there are issues with planning, production capacity, or the shortage of materials, whereas shorter lead times may point to a more efficient supply chain and better planning on the part of the manufacturer.

As an example, if you see that something such as batteries take 3 days to make whereas your competitors need 2 weeks, you have a competitive advantage over them because they are not utilizing their full capacity.

Demand forecasting

It is important for companies who produce consumer goods such as foodstuffs and apparel items to maintain adequate stockpiles of their products at all times so that they will always have something ready for sale when consumers come looking for them.

However, this requires having large inventories relative to demand at any given time, which means being able to accurately predict demand is critical in order to keep costs under control while still maintaining adequate levels of inventory to hand.

AI is a powerful tool for forecasting as it can be trained on vast amounts of historical data and find patterns that influence demand automatically.

Maintaining adequate cash flow

Maintaining sufficient levels of available cash flow during periods where sales volumes are low helps ensure there is enough liquidity available within a business so that it doesn't need external financing. Having too much debt within a business could create problems down the road if revenues suddenly fall below what was originally budgeted due to changes in spending patterns among customers (for example, people buying fewer pairs of shoes than expected).

On the other hand, having too little cash flow during slower periods may mean delaying investments or purchases necessary for future growth until economic conditions improve; this could mean missing out on market opportunities if those investments had started earlier.

Just as AI can be used for accurate demand forecasting, given historical cash flow data, consumer goods brands can forecast cash flow and make changes accordingly. Better inventory management plays a role as well, as overages and stockouts negatively impact cash flow.

Analyzing the impact of discounts

In many consumer goods industries, consumers can often get deep discounts on their purchases throughout the year thanks to promotions that retailers offer. Discounts can also be offered to entice new customers who may not have been aware of such deals before becoming regular buyers at that retailer.

Discounts directly impact both gross margins and net margins for companies that sell similar products since they effectively lower prices. So, while discounts do help boost sales in the short term by increasing volume, they should not be relied on by companies for long-term planning since they tend to eventually wear off.

With AI, consumer goods brands can predict the impact of discounts on any given product or product line, both now and in the future.

Identifying seasonal trends

Seasonal fluctuations are very common within most industries, as well as specific sectors within those industries. In some cases, these fluctuations result from cultural preferences regarding holidays or other events, such as sports tournaments that may draw large crowds for a period of time.

Comparing current sales figures with past years' figures can give a company insight into which periods were best. This information can then be used to help determine whether current expectations for upcoming months are likely to meet or beat those previous expectations.

Social media analytics

Social media platforms offer businesses many ways to interact with their target audiences – including the ability for companies to distribute product information via posts made by employees or contractors who work remotely but still interact with followers online (for example, via **Twitter** posts about new products being added to product catalogs).

This kind of interaction between employees and followers offers another opportunity for brands looking to engage with their target audiences, even when they aren't physically present in person – and social media platforms make this easy for companies who utilize them effectively.

Further, consumers are constantly talking about products and brands on social media. With AI, brands can more effectively conduct **social listening**, and they can analyze consumer sentiment and wish lists from social media, in real time and at scale.

Now that we understand how to analyze product data across a range of settings, let's look more closely at using Commerce.AI for consumer goods brands.

Using Commerce.AI for consumer goods brands

As we now know, consumer goods brands are facing greater challenges than ever, and the solutions to these challenges can be found in the data. However, data alone is not enough, and consumer goods brands need to find a way to turn that data into actionable insights.

Let's explore five main ways in which Commerce.AI can be used for consumer goods brands by gaining value from that data:

- Measuring product attributes and trends
- Predicting revenue opportunities
- Analyzing user personas and customer segments
- Analyzing the customer journey
- Generating consumer goods product ideas

Let's explore each of these in the following sections.

Measuring product attributes and trends

Product innovation is the best way to grow a consumer goods business. However, many managers struggle with understanding what makes their products different and better, which can result in missed opportunities.

The reasons for this situation are complex, but they include barriers such as complexity of measurement, limited datasets being available for analysis, a lack of domain expertise required to understand patterns, and a lack of reliable metrics that can be applied across multiple product lines. These challenges have made it difficult for companies to effectively develop new products and services.

Let's focus on one particular dimension of Commerce.AI – sentiment analysis associated with product attributes – through the lens of innovation in consumer goods. One intriguing finding from our research is that attributes such as taste or brand expectations play an important role in shaping purchase behavior behind consumer goods brands.

For example, one reason why upscale consumers buy **Prada** is that they expect a higher price tag alongside an interesting design aesthetic rather than the standard knockoff experience you'd expect from buying a bag off the discount rack at a department store.

Similarly, one reason **Starbucks** has been so successful is not just its focus on premium coffee but also the expectation among its customers that it will create interesting new flavors through its roasting technology, rather than it rehashing old favorites time after time. The same logic applies when looking at technology firms versus commodity vendors – there's something interesting happening here that warrants further investigation.

When you strip out all the noise, there are two types of meaningful signals emerging in product data:

- Signals about underlying quality and perceived value creation being performed by CPGs versus their competitors
- Signals about how likely customers are to pay certain prices

In other words, purchase intent ripples through both product attributes and attributes that lead to higher price points, such as service levels or technology features. In short, AI offers real promise in this context by augmenting traditional market surveillance capabilities.

Figure 8.4 visualizes a mockup of Commerce.AI for measuring product attributes and trends, which it fills with the relevant brand and product data:

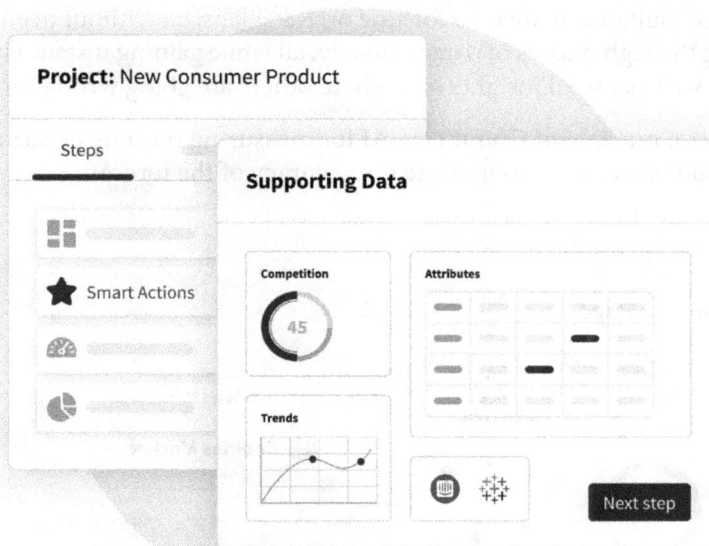

Figure 8.4 – A Commerce.AI mockup for measuring product attributes and trends

While it's great to understand product attributes and trends, revenue is an undoubtedly crucial measure to consider as well. Let's take a closer look at measuring revenue opportunity.

Predicting revenue opportunity

The art and science of product innovation have always been about finding new ways to satisfy customer needs and desires – and generating revenue along the way. Disruptive innovations often succeed by taking market share away from depleted or disrupted incumbents. Such disruption requires an ability to identify opportunities by their revenue opportunity.

This can help uncover new growth drivers, discover business models that aren't obvious in retrospect, help you to prioritize investments in uncertain areas, and provide early warning when competitors enter an untested space (which could put you at risk of industry disruption if you don't react quickly enough with superior offerings).

With AI as the enabling technology, it is possible now for organizations of all sizes to gain this type of competitive advantage over their peers – all with relative ease; all without having to hire expensive consultants or spend a fortune on R&D labs; all without needing prior experience mining through masses of data manually; all while gaining instant visibility into the path ahead, as well as useful insights into where others are going wrong (or right).

Figure 8.5 visualizes a mockup of Commerce.AI for measuring revenue opportunity, which includes a *confidence score* to indicate the accuracy of the forecast:

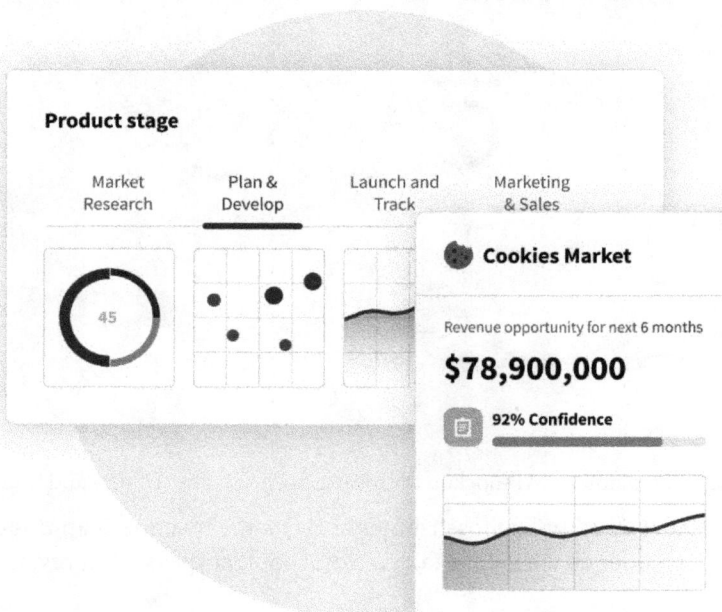

Figure 8.5 – A Commerce.AI mockup for measuring revenue opportunity

Once the revenue opportunity is understood, it's time to better understand user personas and customer segments within that market.

Analyzing user personas and customer segments

Segmentation is an essential aspect of the product-market fit. And it's something that simply can't be ignored by entrepreneurs and product managers.

That said, knowing when to segment users (or potential users) and build features around them can be tricky. The key here is finding the right balance, which requires understanding your user personas as well as your customer segments.

So, what should you know before segmenting your customers? What are the best approaches for building features around user personas, and what pitfalls can you avoid? Read on for tips on how to make this seemingly difficult process easier.

At their core, **user personas** are a simple exercise in empathy. You simply try to understand who exactly uses your product and what they're trying to accomplish with it. By doing this, you'll be able to build features – or at least concepts for building features – that resonate with the users you already have, rather than seeking out new users that might not even exist yet.

Figure 8.6 visualizes a mockup of Commerce.AI for measuring personas and segments, which it fills with the relevant consumer data:

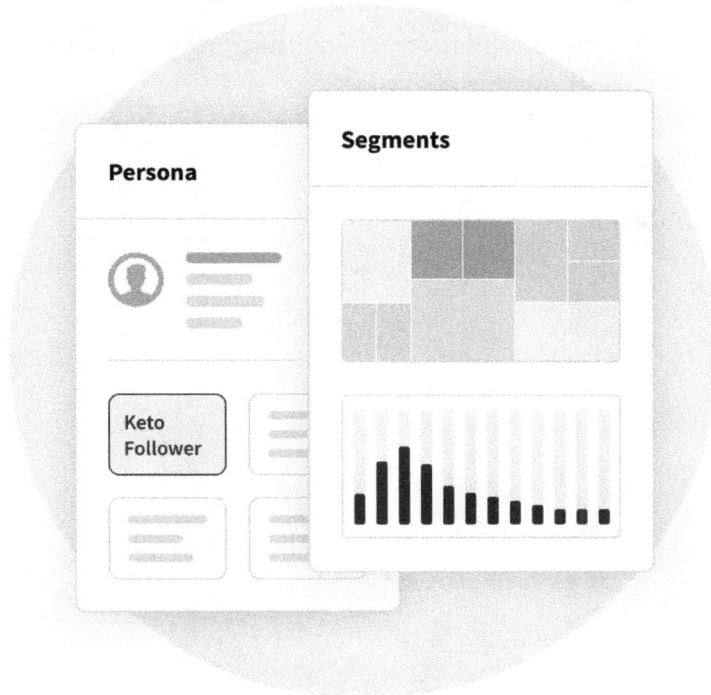

Figure 8.6 – A Commerce.AI mockup for analyzing personas and segments

Understanding user personas and segments is a critical part of gaining a better understanding of the market, but we also need to take it to the next level by analyzing the customer journey.

Analyzing the customer journey

The **customer journey** is the path that a customer takes from first learning about your product to becoming an active user of it. Understanding this journey and how to influence it is critical for driving innovation, increasing engagement, and reducing churn.

Traditional analytics methods focus on measuring changes in key metrics such as revenue, user engagement, or **cost of customer acquisition** (**CAC**). While these are useful metrics to track over time, they often fail to tell the whole story. In particular, CAC fails to account for the value of having engaged users (that is, those who remain active every day) further down the line.

That's why we need new ways of looking at customer behavior. Using AI allows us to identify patterns and make inferences that would be too complex or costly for any human analyst to produce. We can then use this insight to inform our strategy and improve our product design and development process.

Figure 8.7 visualizes a mockup of Commerce.AI for measuring the customer journey:

Figure 8.7 – A Commerce.AI mockup for analyzing customer journeys

Once we've gained this deep understanding of the market and its potential customers, it's time to come up with product ideas.

Generating consumer goods product ideas

In *Chapter 6, Applying AI for Innovation – Consumer Electronics Deep-Dive*, we used large language models to generate consumer electronics product ideas with Commerce.AI. Here, we can use the same techniques to generate consumer goods product ideas, which can help product teams speed up the brainstorming process.

Using this approach, we've been able to generate thousands of unique product ideas with little effort. For example, in *Figure 8.8*, we can see three product ideas generated around *healthy snacks*:

GENERATED IDEAS

	INDEX	SCORE	SAVE	PRODUCT IDEA
	10	★ ★ ★ ★ ★	Copy	A high-protein, healthy alternative to traditional chocolate peanut butter bars, with the same creamy and crunchy consistency.
	9	★ ★ ★ ★ ★	Copy	A healthy oils potato chip without vegetable oils.
	8	★ ★ ★ ★ ★	Copy	A no-sugar, healthy-fat based yogurt.

Figure 8.8 – AI-generated product ideas around healthy snacks

As we've seen, Commere.AI can be used to measure product attributes and trends, predict revenue opportunities, analyze customer data, and generate product ideas. This is a powerful tool to speed up the product development life cycle, which is key to getting ahead of your competition and being first to market.

Summary

In this chapter, we've learned about the key challenges facing consumer goods brands, and how to overcome them by analyzing product data and using the Commerce.AI data engine.

Data can be used to overcome significant challenges in the consumer goods space, which is why many leading brands and product teams are investing in AI to gain an advantage.

In particular, we've learned how to use AI for tasks such as consumer goods content generation, analyzing consumer goods reviews, and demand forecasting. We've also looked at how consumer goods product teams can use Commerce.AI to measure product attributes and trends, predict revenue opportunities, analyze user personas, and more. In doing so, consumer goods brands can better innovate and launch successful products.

In the next chapter, we'll explore in detail how Commerce.AI's **Product AI** can be used for product concept and development, product launches, and product management. These insights will be useful for any product firm.

Section 3:
How to Use Commerce. AI for Product Ideation, Trend Analysis, and Predictions

In this final section, you will learn how to leverage Commerce.AI's data engine to ideate, predict, and analyze new product ideas, trends, and demand for products across categories.

This section comprises the following chapters:

- Chapter 9, *Delivering Insights with Product AI*
- Chapter 10, *Delivering Insights with Service AI*
- Chapter 11, *Delivering Insights with Market AI*
- Chapter 12, *Delivering Insights with Voice Surveys*

9

Delivering Insights with Product AI

Data is a goldmine for product teams, allowing you to develop better, more targeted offerings for your customers. You can use data to build actionable insights into how people interact with your products and make decisions that will help you deliver the best possible product experience.

But simply *having data* isn't enough. What matters is *how you use it*. The key, as we'll discuss in this chapter, is to take a holistic approach that integrates data into the entire product innovation life cycle. In this chapter, you'll learn about how to integrate data into the product innovation life cycle, including decisions around product development, product launches, and more. In short, you'll learn how to use data to bring better products to market, faster.

We'll discuss how to use **Commerce.AI**'s **Product AI** features for every stage of the product life cycle, from market research and product ideation and creation all the way to post-launch ads and sales. In particular, we'll cover the following topics:

- Commerce.AI for product concept and development
- Commerce.AI for product launches
- Commerce.AI for product management

Commerce.AI for product concept and development

Product conception is an old art, but the underlying principles are still valid. The main principle is that people buy for a reason—they want to solve a problem or meet an unmet need.

After all, guesswork will inevitably lead to a product that consumers don't want or need, which will fail in the market. Therefore, product teams need to first conduct extensive market research in order to understand consumer wants and needs before the **product ideation process**.

Let's explore the following areas within product concept and development in the next subsections:

- Market research
- Understanding demand
- Product ideation

Market research

The product development workflow starts with market research, followed by concept ideation, design, and engineering. This is how most products are launched into the market. But in a world of on-demand AI solutions that can provide data insights from inception, many companies have begun to use AI to shorten the **product development life cycle** (**PDLC**). By applying AI at each step of the PDLC, you can make sure your products aren't just informed by data but actually reflect it throughout the entire process.

Market research for product teams is all about understanding the needs of customers as they engage with a product or service. It's about finding out what customers really want, and it's about figuring out whether a new idea will work.

Market research goes beyond simply understanding what people currently think about an idea or product. Marketers also need to understand how people will change their minds in the future, and why. And market research is not just for product teams; all teams at any stage of product development (marketing, design, engineering) should consider market research a critical part of the process.

Further, market research is one of the most cost-effective ways to develop product ideas, especially if you do a lot of user testing to validate your hypotheses. The cost of gathering some key data points can save you months of development time when compared to trying to design things blindly. That said, traditional market research has its limitations.

Limitations of traditional market research

Traditional consumer surveys are tedious, time-consuming, and cost a lot of money, and they often prove disappointing when it comes to product innovation challenges. Moreover, consumer preferences change quite frequently. Traditional research methods are hardly equipped to capture the complexity of users' motivations.

With AI in commerce, you can update your process for market research to provide exciting new avenues for product innovation. You can supplement qualitative market research with automated quantitative market research, which can provide the data insights you need to help validate product concepts.

With Commerce.AI's AI-powered market research reports (`https://www.commerce.ai/reports`), you can streamline the research process with low-cost, online reports to discover exact insights about customers and the sizes of market opportunities.

Leading product firms are applying AI to market research to improve product development. After all, *speed to market* is a competitive advantage, and keeping up with customer input and feedback is key in this.

Understanding demand

Understanding demand through market signals is one of the most important steps in product innovation. For decades, market research has relied on a combination of quantitative and qualitative methods for identifying trends in consumer behavior that can be used to inform your product strategy and design.

There are many different types of market signals that marketers can observe, but two major categories include quantitative and qualitative observations:

- Quantitative signals include things such as *search volume* or *sales velocity*.
- Qualitative observations involve *talking to customers*—both directly (via voice surveys) and indirectly (via online forums, social media, or product review replies).

Simply put, the digital space offers new ways to observe how people are using products and services—and in fact, there's no shortage of data available now about what people do when they interact with technology. Businesses can collect this data, and it can be of great value.

The challenge, however, is that the data itself is not market research—it's a *starting point* for market research. Marketers need to transform this raw information into insights about what people want, how they want it, and why they want it, in order to understand what will result in demand.

By understanding demand, companies can go beyond simply creating a product or service that someone wants and start thinking about how to create a product or service that people don't even know they need yet (but will end up loving nonetheless).

In *Figure 9.1,* we can see a mockup of Commerce.AI's **Market signals** dashboard, which it fills with relevant product data from a variety of sources for the brand using it:

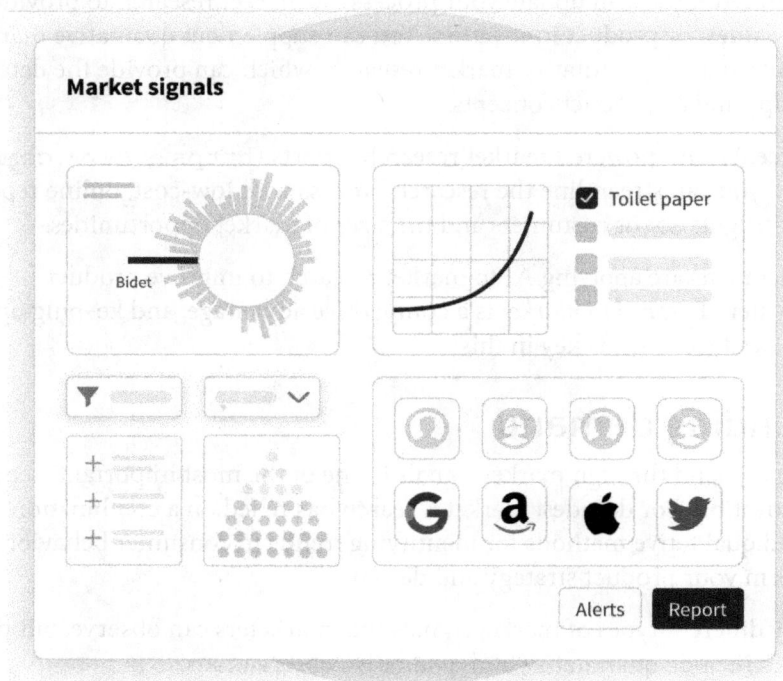

Figure 9.1 – A mockup of the Commerce.AI Market signals dashboard

As seen in *Figure 9.1,* consumer goods firms discovered during the COVID-19 pandemic that bidets and toilet paper surged in demand, which signaled an opportunity to create or enter into a new category of hygiene products. With Commerce.AI market signals, brands can also set alerts to notify them about important events, such as a spike in demand or interest in a particular product or service.

The key to making this process work is using data and insights to make informed decisions about what will result in demand. The more data and insights you have, the better you can create a roadmap for your product or service, which will help guide you as you move from concept to launch. And the better your product roadmap, the better your product or service will ultimately perform.

Product ideation

After exploring demand opportunities and market signals, it's time to start thinking about your next steps. One of the next stages in the PDLC is *concept ideation*. In the past, you might have drawn on your own experience for ideas. But with data, that's a thing of the past. With AI, you now have access to user data at scale to provide context for new products.

Let's look at two main ways in which to use AI for product ideation. First off, AI fueled by product data can uncover what features and attributes your competitors are already providing to consumers. These insights can help you ensure that your next product release doesn't overlap with an existing one. This enables you to enter new product markets with more competitive differentiation. It can also extract patterns and trends from the data to help you understand what consumers want and need.

For example, let's suppose you're trying to expand your product line of electronics accessories and sell a cell phone charger. As seen in *Figure 9.2*, we can view an AI-generated feed of leading brands and products in this area. In particular, **Anker** is the brand, and its three main strengths are that it's *fast*, it holds a *long charge*, and it's *durable*. Therefore, we know that to compete with Anker, those are the top attributes that we'd have to provide to consumers as well.

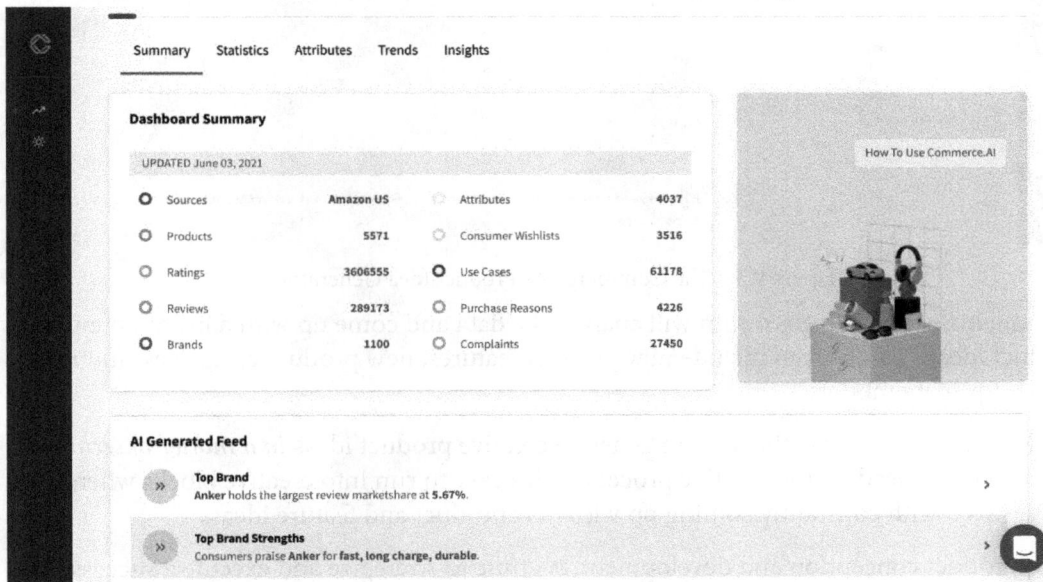

Figure 9.2 – Commerce.AI Dashboard Summary for wall chargers

The second way AI can help you is by generating new product ideas. AI can generate any number of product ideas in a matter of seconds, and this can be a huge time saver.

Let's look at exactly how this works. The data used to train the algorithm can come from a variety of sources, including your own product data, market data, and user research data.

For example, in *Figure 9.3*, we use the **Commerce.AI Product Idea Generator** to create a new idea for a cell phone charger. The idea is *a high-speed charger that is small enough to fit in my pocket*. This generated idea makes sense, as typical chargers are often very bulky and inconvenient. Perhaps a product team will go along with this idea, or perhaps it will spur another idea for a different kind of product.

Figure 9.3 – The Commerce.AI Product Idea Generator

The **machine learning** algorithm will analyze the data and come up with a list of potential product ideas. This list can include new product features, new product categories, and new product pricing models.

The key takeaway here is that AI can generate creative product ideas *in a matter of seconds*. This can help speed up the creative process, as it's easy to run into creative blocks when you're faced with constantly coming up with new product and feature ideas.

After product conception and development, it's time to strategize and execute a successful product launch.

Product launch

In the past, product managers had to rely on gut feelings and their own experiences to make the right product decisions. Today, AI is changing this dynamic by providing a more objective way of making product decisions based on data and past interactions.

Let's take a step back and look at how AI is changing the world of products, with a focus on four areas:

- How AI is changing product launches
- Predicting demand from early signals
- AI for the two types of product launches
- Using AI for product launches—advantages and disadvantages

Let's start off with an overview of how AI is changing product launches.

How AI is changing product launches

Traditional product innovation was built around the *interview-centric* approach. In other words, product managers assumed that the best way to understand their customers was to spend time with them. This approach has evolved quite a bit over the past few decades. The trend now is toward more data-driven approaches to product management.

In the last decade, the hype around AI was growing rapidly. The concept of using software to automate repetitive tasks seemed like a no-brainer and something every company needed to be doing if they wanted to stay competitive in an increasingly fast-paced business environment.

However, assembling the right team and building an AI product from scratch was quite an undertaking back then. In fact, just getting access to good datasets took significant work. Companies either had to build their own, or they had to spend huge sums acquiring them from competitors.

With Commerce.AI, this treasure trove of product data is readily available to any company that wants it. The idea behind AI-powered product launches is simple—automate as much work as possible, and create an environment where humans can focus on innovation and creating something truly unique for customers.

The end result is a faster time to market for new products with fewer risks along the way. This also means freeing up valuable resources for other initiatives within your company. So, how exactly does AI help with product launches? One way is by predicting demand from early signals. Let's explore that concept in detail.

Predicting demand from early signals

The first thing to realize is that AI can't be relied on 100% of the time. It doesn't always work, and it has its limitations. However, when used properly, AI can provide invaluable insights into customer behavior that can help companies make informed decisions about product launches while reducing risk along the way.

This type of approach is called **pro-active demand analysis** because it uses machine learning techniques to predict demand before it arises (rather than reactively, during times when sales are slow or zero). By doing so, companies can also ensure that they have enough inventory on hand to meet peak demand after a product launch—which could result in lost sales if an item isn't restocked quickly enough.

This concept goes beyond product launches—anything where inventory management plays a role should be improved with AI capabilities. After all, most commerce businesses need to constantly replenish the stock of any given SKU.

The key takeaway is that these types of machines learn patterns and thus enable humans to do high-level tasks, such as anticipating changes in consumer behavior and taking action before something becomes an issue, rather than dealing with issues once they arise.

AI for the two types of product launches

To better understand the challenges and opportunities companies face with launching new products, it's helpful to see how these issues play out in different environments. In particular, we need to pay attention to two main types of product launches:

- Hard launch
- Soft launch

Let's first explore the concept of a hard launch and how AI can help optimize this event.

Hard launch

A **hard launch** is a full-on, company-wide campaign that is designed to introduce an entirely new product or service to the marketplace. Typically, this involves extensive media buys, advertising campaigns, and events—all designed with one goal in mind: getting as many people as possible to try your product for the first time. The hope is that once they do, they will become brand advocates who will tell their friends and family about your product.

With Commerce.AI, you can use AI to optimize hard launches by automating much of the communication and content creation process. You can also use the toolset to plan ahead and prepare in advance of a big launch to ensure your content is optimized for success.

Additionally, you can segment your audience to ensure that each person is getting the most relevant content so they have the best chance of engaging with your company. You'll also be able to analyze the performance of your marketing campaigns right after the launch, seeing not only how many people viewed your content but also which pieces performed the best.

Soft launch

A **soft launch** is a controlled, limited release of your new product or service to a select group of people who have been handpicked by you, the company. The goal here is to gauge customer reactions and measure **customer acquisition costs** (**CACs**). In other words, you are testing out your product or service to determine how much it will cost to acquire customers, and then adjusting your launch strategy accordingly.

Typically, this involves sending an email newsletter announcing the soft launch to a specific list of people, before opening it up for anyone who wants access. You can also do a soft launch on social media—just announce that your new product is available for early access and only invite select people via invitation emails. You can test different price points as well as different types of content in order to determine what works best on each platform.

Even if you don't get enough interest during the soft launch period, if you use analytics tools such as Commerce.AI, you can see which messages resonated with customers and adjust future campaigns accordingly.

For example, if a particular message brought in more sign-ups at a higher average revenue per user than others did, that would be an indication that this message would work well as part of a larger campaign later down the road (once you had collected more data about customer behavior from real users).

Voice surveys are another way to gain feedback from any kind of launch. Unlike traditional text-based surveys, voice surveys have particularly high engagement and completion rates, and they can be used to gain valuable insights after a launch. In *Chapter 12, Voice Surveys*, we'll discuss their use in detail.

Using AI for product launches—advantages and disadvantages

In addition to helping companies avoid potential supply chain disruptions, pro-active demand analysis has numerous advantages as well. It helps eliminate guesswork regarding what products may sell better based on previous experience with similar products. It also enables companies to avoid under/overstocking based on potential demand, which can lead to greater efficiency and profitability, as well as less inventory risk.

When it comes to AI in the context of product launches, this leads to lower costs and more streamlined processes overall. Additionally, AI helps companies focus their resources on creating something truly unique for customers, rather than spending time and resources building out infrastructure that they might not need.

Using AI will free up team members who are better suited for other tasks, such as managing growth or developing new products altogether. While a successful product launch is key, that's far from the end of it, so let's explore how to best use data and AI to improve post-launch product management.

Product management

It's fair to say that many product managers have over 100 tabs open in their browsers at any given time. But what if there was a single tool or service that could monitor all of your products, across all channels?

And what if you could use it to get actionable insights into how your products were performing in the market? And what if this information was available on an ongoing basis (real-time), so you always knew where your product stood and what the trends were? With Commerce.AI, you can monitor your products in the market and across channels.

The core principle behind AI in product management is to leverage machine learning and AI techniques to give you actionable insights into how your products are performing. By using AI, you'll be able to make better decisions about what changes should be made to improve your product's performance, or even pivot and change the direction of your product entirely.

In this section, we'll look at AI for product management across six areas:

- Tracking product wishes
- Brand management
- Using AI for consumer insights
- Using AI for product tracking
- Marketing and merchandising
- Customer support

Let's get started by looking at how and why product managers can use AI to track product wishes.

Tracking product wishes

One significant use of AI in product management is **sentiment analysis**. Sentiment analysis uses **natural language processing** (**NLP**) techniques to analyze written text for signs of positive or negative sentiment. Sentiment analysis can reveal whether users think a particular feature of a product is useful—that is, it can tell you what people like or dislike about your product. In other words, sentiment analysis provides information on how people talk about your products—good and bad!

As a result, product managers can use AI to track product wishes, both in terms of features to improve or remove, and features that consumers want to see added to a product.

Tracking a product wish list is an important exercise for any product manager, but it can be challenging to maintain the data and have a system that you can refer back to throughout the life cycle of a product. This can result in a dilution of focus as well as losing track of key considerations, such as the following:

- What are the top features or use cases desired by your customers?
- How does your product stack up against these use cases or wishes?
- Are you meeting them, exceeding them, under-delivering, or falling short?
- What else do you need to know about this topic in order to build a great product?

Commerce.AI offers real-time tracking on all your products across channels, and it also offers historical insights for you to have context over time. With Commerce.AI, you can set parameters on what changes matter most when building out new features, test new concepts by enabling experiments on a subset of users and measure their engagement, and understand how these new ideas impact revenue or other **key performance indicators** (**KPIs**). Then, you can make informed decisions about which ideas to prioritize and invest resources into moving forward.

All this is possible from within one tool—no third-party software is required. We believe this kind of holistic insight into product performance will become even more critical for many organizations in the years ahead.

Brand management

Product management and brand management go together like hand and glove—they're the core of a product company. Without branding, we wouldn't know what to make of a company's offerings, how trustworthy they are, or what their target market is. With the rise of AI and the integration of machine learning into product management tasks, it's now easier than ever to successfully use big data to manage brands.

Companies are now able to leverage AI to gain a better understanding of who their customers are, how they behave, what they like, and where they are. Commerce.AI can be used for a blend of brand and product sentiment analysis, consumer insights, and product tracking in order to identify opportunities to create and maintain better brand experiences.

Using AI for brand sentiment

Commerce.AI's brand sentiment tools have been built on the NLP stack, which means that we leverage techniques such as machine learning and deep learning to analyze spoken language in natural contexts (for example, from customer service interactions). This allows us to automatically extract information from unstructured data, such as social media posts or comments on product review videos.

This technology gives us the ability to identify either positive or negative sentiment about brands in social media posts and comments. Sentiment analysis is a well-known field of study within NLP. Software designed for this purpose is now widely available, with many tools being developed over the past decade.

In general, there are several categories of sentiment analysis algorithms:

- **Word count algorithms** (which count the number of times certain words occur)
- **Co-occurrence algorithms** (which look at how often specific words occur together)
- **Polarity/agreement algorithms** (which determine whether a post expresses more positive or negative sentiment)

By using these well-established technologies to mine public conversations online combined with deep neural networks, we can gain insights into what people think of various brands—both good and bad—helping you to understand your customers better and devise strategies for promoting your brand in a way that will play to its strengths while avoiding potential weaknesses.

By using machine learning techniques to mine public conversations, you have an opportunity to read between the lines without having to talk directly with customers face-to-face or go through lengthy market research studies.

Next, let's look at consumer insights through the lens of product management.

Using AI for consumer insights

One area where AI is especially well suited to product managers is consumer insights. As a brand manager, you can use machine learning and NLP to identify the types of data that will be most relevant to your particular business objectives, and then you can automate the collection and processing of this type of data as part of an ongoing effort to stay abreast of your target market.

For example, if you're developing a new mobile app, you may want first to look into some basic demographic information about your users (that is, age, gender, location, and so on) in order to understand more about them as customers before building out additional features that will appeal to them specifically as individuals.

Likewise, if you're a retailer trying to develop a new e-commerce site or chatbot experience for your customers, knowing what kind of products they are looking for right now can help inform what you build in the future.

This is also true if you're building a brand new website or mobile app. You want the site or app development process itself to be customer-centric from the ground up so that it's always focused on understanding who your users will be. With AI, you can more deeply understand your customers and therefore design more tailored experiences for them.

Now let's look at another important aspect of managing brands using AI—product tracking.

Using AI for product tracking

Managing product life cycles isn't something that should be done manually anymore— there's too much detail involved with validating all the different touchpoints within each stage of the product life cycle, all the way from idea generation and development through to testing and launch.

Fortunately, AI can help streamline this process considerably in order to get products out faster while maintaining high standards. If properly implemented across multiple teams within an organization (for example, product management, engineering, operations), automation efforts can also lead to increased accountability among team members.

Product managers are tasked with ensuring that a company's offerings meet customer needs while staying on time, on budget, and within scope. And like many roles within companies today, this often involves juggling multiple tasks—managing stakeholders across multiple departments, keeping abreast of competitive developments in the industry, understanding customer behavior patterns—it's no wonder that coordinating these efforts is a full-time job!

However, integrating AI into your workflow can free up valuable time so you can focus on more strategic activities that will benefit your organization.

Marketing and merchandising

When it comes to merchandising, an obvious step to take with AI is selecting the *right products* to display at the *right time*. In fact, this may well be the quickest win you can get from AI.

The key here is to apply AI in a way that's actionable for your business. For instance, if you're a retailer, applying AI to product selection could mean identifying bestsellers among similar items (for example, which sneakers/boots are selling well?) and using **A/B testing** or other tools to determine what specific attributes of those products make them *sticky*, and pushing more of those kinds of items onto shoppers while curating less popular items off the site altogether.

A/B testing is an effective way to measure the impact of specific product attributes on customer behavior, and it's often used by e-commerce companies to optimize their site for increased conversions.

The letters *A* and *B* in *A/B testing* refer to *control* and *treatment*, respectively. The control group is the original, while the treatment group receives a specific change. Suppose you were testing a new pair of sneakers on **Amazon**. You might change the default color from black to red or raise the price by 10%. You'd then measure how many people click on that specific item, compared to how many clicked on the original sneakers.

For example, if you're an online fashion retailer, using AI for product catalog optimization could mean identifying top sellers based on attributes such as popularity and price-to-value ratio and then using data science to optimize what goes into your product catalog so that it matches shopper needs as closely as possible.

With the explosion of online data sources at our fingertips, we can now analyze previously uncalculated data points, such as traffic trends over time or how different types of shoppers interact with your website compared to others. We can also mine that mountain of data for patterns and use them to inform better decisions across multiple channels (such as which personality types might respond best to different campaign content).

It's also important to take advantage of customer habits. Customers now expect a seamless shopping experience across various channels, and they're willing to engage with brands in new ways. With that in mind, it's worth considering which shopping behaviors have the highest value for your brand and why. Some key questions to ask yourself include the following:

- How does product interest vary by day of the week?
- What types of purchases are shoppers most likely to make when they're planning a vacation?
- Are there any other shopper personas you should be paying attention to?

Besides taking advantage of customer habits, it's important to empower consumers through discovery. When customers are seeing products and services they're interested in, they'll stick around longer. So, using AI, you can help them discover relevant content based on related searchers and past purchase history.

And if that content doesn't exist yet, you can create it. It's never been easier to develop new content and even new product ideas with AI. The technology is getting better at understanding what consumers want and need, so you can create new types of content and product experiences that didn't exist only a few years ago.

Product copy and packaging

There are many ingredients to successful marketing and merchandising beyond the product itself, including ad copy, product descriptions, and product packaging.

In *Figure 9.4*, we can see how a mockup of Commerce.AI can be used to automatically generate product copy for product catalogs, listings, ads, **Instagram** posts, and more:

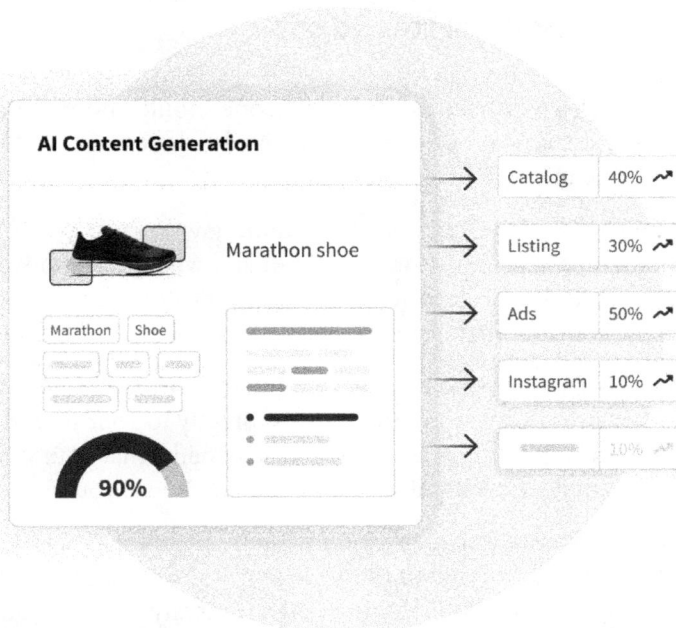

Figure 9.4 – A mockup showing Commerce.AI's AI Content Generation features

The first ingredient, *ad copy*, is a means to convey your brand message while establishing credibility and trust. It's how you tell people what your product or service can do for them. When it comes to ad copy, it's all about the benefits you communicate in an effort to unlock the value of your offering through different channels (AdWords being one example).

Another ingredient in successful products is the product description. The product description is how you tell people about the product and why they need it. It's a means to help people understand what your product can do for them and how it will benefit their lives. The key here is to communicate the benefits of your offering in terms that are relevant to your target audience so that they can understand its value proposition.

Lastly, we have product packaging, which is a means to help people see and experience your product or service before they buy it. It's all about making sure that the package itself conveys the right message so that when someone opens up the box or unwraps the item, they get a sense of what you want them to feel or experience with your product. This helps create an emotional connection between you and your customer so that when they make a purchase decision, they choose to buy from you instead of someone else.

So, these are some of the ways in which you can use AI for marketing and merchandising. Now, let's dive into how AI can be used for generating ad copy and product descriptions in greater detail.

How AI can help you write better ad copy

Ad copy is critical because it helps convey who you are as a business, what makes you different from everyone else out there, and why someone should choose to do business with you over another company or option available. And if done right, this will result in more conversions (people buying).

The first step in writing effective ad copy is understanding who exactly will be reading it (the person who sees it on social media or elsewhere) and what their motivations might be for doing so. Once we know this information, we can start crafting our ad copy around these insights in order to ensure that our message resonates with our intended audience.

The second step in writing effective ad copy is defining the core benefit or promise of your product or service so that you can communicate this directly to your intended audience. This is where we start crafting our message around the benefits that your offering provides for people's lives so that they understand why they should choose you over someone else.

With AI, you can scan a mountain of data around your consumers and other competing products to understand exactly what kind of messaging you should create. Similarly, you can use large language models in Commerce.AI to actually automatically generate that copy.

How AI can help you write better product descriptions

The next ingredient in successful products is having an effective set of *descriptions* on each item that goes along with it.

Just as AI can be used to generate ad copy, the same models can be used to generate product descriptions, which are similar pieces of text designed to convert viewers into buyers. Unlike ad copy, you have a lot more room for text and explanations in product descriptions, which also play an important role when it comes to **search engine optimization** (**SEO**).

By using AI to generate product descriptions for all your products, you can ensure that consumers find your products, effectively boosting sales.

Customer support

Product teams are increasingly using AI to help them build better products. But there's a disconnect between the insights that product teams can gain from AI and how they use them to support their customers. In this section, we'll explore how product teams can use AI for customer support.

The first step is understanding what *customer support* means in the context of product development. Customer support is about helping your users solve problems with your product or service. It's about anticipating issues before they happen and providing actionable advice when they do. It's about understanding what makes people tick so you can anticipate their needs and provide them with the right information at the right time. And it's about proactively addressing non-issues so that people don't have to call you for help in the first place!

In other words, customer support is all about empathy—understanding your users and anticipating their needs so you can provide them with relevant information at just the right time, whether that's during their use of your product or service or after a purchase has been made. This requires an intimate knowledge of your users, which you can gain by asking the following questions:

- Who are they?
- Where do they come from?
- What are their goals?
- What are their pain points?
- How do these intersect with your own business goals?

And most importantly—how can you best serve them by leveraging insights around product issues, comparative advantage, and use cases to empower your interactions wherever you have them?

In this way, customer support and product innovation are closely intertwined. Therefore, let's look more specifically at how product teams can use AI for customer support.

How product teams can use AI for customer support

Product teams can use AI to stay ahead of their customers and leverage insights around product issues.

Product teams rely on data to inform decisions around design, engineering, marketing campaigns, partnerships, and more, but data alone isn't enough—it must be actionable data that helps drive results for the business. Data without insight is noise; insight without action is blind speculation; action without accountability leads to missed targets, or worse yet—no targets at all!

So, how does a team stay accountable when using analytics as part of its decision-making process? The answer lies in applying critical thinking skills to the data. When you apply critical thinking skills to your data, you're able to see patterns and draw inferences that can lead to insights.

With critical thinking, you start to understand what is happening in your product or service, why it's happening, and what you can do about it. This requires a shift from purely analytical thinking (which focuses on the numbers and metrics) toward a more holistic approach that considers context and **user experience** (**UX**).By applying critical thinking skills to your data, you're able to see patterns and draw inferences that can lead to insights.

Insightful questions product teams should be asking themselves include the following:

- What are the most common customer issues?
- What are the top reasons for customers churning?
- What are the most commonly reported bugs?
- How does our product compare with similar products in terms of usability, accessibility, and more?

These types of questions require an understanding of how people interact with your product so you can make informed decisions about design, development, marketing campaigns, and partnerships, based on actual user behavior rather than just assumptions or best-guess estimates.

It also allows them to take action earlier in the process, so they aren't left scrambling at the last minute and trying to fix things after a release has gone live.

For example, a team may have identified from online data sources that many of its users are having trouble signing up for an account. The team could use AI tools such as chatbots or **virtual assistants** (**VAs**) as part of its support strategy so that users don't have to call customer support for help setting up their accounts—instead, they can simply message their VAs, who will guide them through the process over video calls or **instant messages** (**IMs**).

In this scenario, using AI would allow the company not only to save money on customer service costs but also provide a better experience for its users by eliminating unnecessary steps from their onboarding process.

Ultimately, product management is a multi-faceted and complex process that spans from tracking the market and managing your brand through to merchandising and customer support. Product managers can use data and AI to improve efficiency and effectiveness in all these areas.

Summary

As we've seen in this chapter, Commerce.AI's Product AI features make it easy for product teams to explore new ideas, gain insights into user behavior, and pivot rapidly. It can also help teams launch products earlier and with a stronger, data-driven approach.

One of the first places teams should use Product AI is during product discovery. Whether you're working at a small, local commerce firm or an international conglomerate, it may be wise to take some time to understand what customers actually need before diving into developing new features.

Using Commerce.AI's Product AI features during this discovery phase will give you valuable insights into customer needs and behavior, and this can inform both your product strategy as well as your feature prioritization once you have advanced your idea.

Once a product is ready for launch, Product AI can be used to optimize its performance. This means you'll be able to test ideas quickly and with minimal effort, which can result in a more informed decision on whether or not to move forward with a particular feature set or design.

Finally, once your product is live, Product AI can provide insights into how it's performing and identify areas that need improvement. These insights are an essential component of any product team's ongoing success, and using Product AI's tools and data will help you avoid common pitfalls in the product life cycle.

While Commerce.AI is a powerful tool for product teams, it also has many uses for service providers. In the next chapter, we'll explore Commerce.AI's **Service AI** features and how they can be used to empower your front line, manage your locations, and enhance your service offerings.

10
Delivering Insights with Service AI

Just as **artificial intelligence** (**AI**) has enabled product teams to enhance the product life cycle, service teams can now leverage AI to enhance the service life cycle.

As we know, it's not enough to provide a great solution to a customer challenge or problem. The customer insight process and the ensuing service design must be central to any organization that wishes to differentiate itself and deliver a higher-value experience to customers. Now, you can use AI to leverage data and gain actionable insights on how to better serve your customers in the future.

It's important to note that AI cannot read minds. However, it can process vast amounts of data and deliver actionable insights. With AI today, service teams can gain an unparalleled understanding of specific elements of the customer journey, and this can lead them to provide a better service.

In this chapter, we'll cover how to analyze data to better understand your customers, stores, and even employees. We'll then explore how to deliver these findings to your teams to gain competitive advantages.

We'll explore the key areas of focus in **Commerce.AI**'s **Service AI**:

- Empowering your front line
- Managing your locations
- Enhancing service offerings

By the end of this chapter, you'll have learned how to better understand customer affinities, purchase reasons, and challenges, and turn your next interactions into great brand experiences. We'll also explore how to compare and monitor customer reviews across locations at scale to optimize your branches, employees, and services quickly. Finally, we'll identify growth areas and opportunities to boost customer loyalty, find new uses for your store, and get a picture of bottlenecks before they escalate.

Empowering your front line

When it comes to empowering your front line, it's all about providing your service teams with meaningful insights about your customers and customer interactions. Let's look at four ways Service AI makes this possible:

- Better understanding customer affinities
- Better understanding purchase reasons
- Better understanding customer challenges
- Turning your next interactions into great brand experiences

Let's start with using AI to better understand customer affinities.

Better understanding customer affinities

By using an AI-powered solution such as Commerce.AI's Service AI, you can identify the product affinities and service preferences of your existing customers. This data can then be leveraged to create a personalized experience that caters to individual customers' needs and interests.

For example, a restaurant could leverage its customer database, including its customer information, such as demographics, past purchases history, preferred dining times (for example, weekend brunch), and so on, to offer late-night delivery services for orders placed on Fridays and early morning delivery services for orders placed on Mondays.

Or another restaurant could leverage their past purchases to determine what items might be popular with people who typically dine later at night or prefer to have their food delivered (such as desserts). Using this information along with other related data points could help the restaurant better segment its customer base and find new market opportunities based on tailor-made offerings that meet people's specific needs – be they food - or non-food-related in nature.

As consumers, we all have different needs when it comes to buying products and services. We each have different preferences about timing (early morning versus night), days of the week (weekend versus weekday), how much notice we need for an order, what delivery method is most convenient (regular mail versus courier), and so on.

But being able to know what customers' preferences are before they go into a store, website, app, or another purchasing medium greatly increases their chances of finding something that fits their needs. It also gives them a chance to see if there might be other options out there that they didn't know existed but would fit their needs better. This could be for anything from food and clothing items to vacations and investments – the list goes on!

The key takeaway for you is that if you can identify your customers' affinities, there are endless possibilities as to what value-added services you can offer them – and from where those offerings could come.

Better understanding purchase reasons

Customers buy services for many different reasons, which go far beyond their financial value. One of the top reasons for people to buy things is to define their social identity. It's not just a matter of self-image either – it's about being part of a group, which creates a sense of belonging.

In some cases, this can be tied to your community – for example, buying local or consuming content that reflects where you live or interests you have. The key here is that most purchases are driven by emotions whose roots are tied to the customer's community. People want to feel connected in order for them to be happy in life – they need those connections and emotional ties.

People often don't do something until someone else has done it first. This means they'll copy others around them, whether they realize it or not. This applies as much to the service side as it does with all other purchases, especially those purchases that consumers perceive as status symbols (such as a premium airline membership) rather than purely functional (such as insurance). These purchases are often about identity more than value – they symbolize who you are.

Our actions demonstrate our values. By dedicating resources to help your team engage with customers, you're demonstrating that as a business, you care about your customers and value them enough to go above and beyond. This can lead to repeat customers and referrals because people trust a brand when they feel valued.

The lesson here is that AI can be a powerful tool for understanding your customers better, and it's a great way to understand what they think about the offerings you provide and why. Using these findings, you can then create strategies that more closely align with your customers' needs and desires. In essence, AI is about removing constraints so that you can grow in ways that weren't possible before, inspiring innovation along the way.

For example, teams could use Commerce.AI to analyze store reviews, social media posts, and other customer feedback to understand what customers are saying about their service. This can help teams better understand the pain points that the customers are facing, which will then inspire service innovation.

Another example is using AI to analyze purchase data to understand why people buy certain products over others. For instance, if you sell insurance, you could use AI to determine which types of insurance policies have the highest retention rates, and then you could use this information to inform your future service development efforts.

Better understanding customer challenges

From the service perspective, there are a few main challenges that your customers may face:

- **Questions about your product or service**: The first challenge is that they may have questions about your product or service. This could be a simple question about the eligibility criteria of a particular offer, or it could be a more complex inquiry, such as, *Can you explain this product feature in detail?*

  When you think about how to solve this problem, you will want to consider the data that your customer has already provided to you. Where possible, you should look at past interactions with customers and leverage that information to help guide future interactions. If there are any patterns that emerge from these interactions, it can save time and effort for both parties involved. It also helps if those patterns align with business goals (such as increasing sales).

- **Providing personal customer interactions**: The second challenge is related to understanding the context of an interaction. In many cases, today's consumers are accustomed to receiving information through digital channels without having to actually talk with another person. As such, providing customers with information in an impersonal way can feel unhelpful for both sides involved. In today's increasingly connected world, it is important for companies offering services, as well as their customers using those services, to become more familiar with each other so that relationships can develop naturally.

With AI, you can gain unprecedented insights into customer challenges, enabling your service teams to empathize with your customers, and therefore providing a more personal and heartfelt experience.

For example, by analyzing the conversations that people have with your service representatives, you can gain insight into the types of questions that customers ask. This information can then be used to train the next generation of customer service agents.

Another example is using AI to search and understand the products that are most in-demand for a particular customer, and then suggesting similar products when they place an order. By understanding the context of an interaction, you can create a more personalized experience for your customers, which will ultimately lead to higher retention rates and greater lifetime value.

Turning your next interactions into great brand experiences

Customer interactions over the past 100 years were driven by utilitarian logic (*get the job done*) without much concern for the human element. This has changed in recent years with attempts to increase human engagement and improve emotional ties between companies and their customers.

With Service AI, brands can better understand their customers and the real needs behind individual customer interactions, while overcoming the huge hurdle of comprehension. Service AI makes it easier to communicate across digital experiences and deliver on outcomes that matter to your customers, whether you're a small business or an enterprise company.

The adoption of Service AI will enable companies to create more human experiences for their customers, and the technology can be used across digital touchpoints, physical stores, or any other brand experience. It's about creating meaningful interactions that drive loyalty.

The concept of emotional connections is a growing focus in the study of customer interactions. These moments can help build trust and loyalty with your customers. And when you have a strong emotional connection with potential customers, you increase brand affinity – that is, the customers are much more likely to buy from you again than a company they have no relationship with.

To give a specific example, suppose a customer loses their wallet and needs to replace it. They could go to a store to buy a new one, but they may not remember the brand or model number of their old wallet. In this case, the customer would be better off just buying a new wallet online – no need to go into the store at all.

But what if that customer is forgetful and prefers in-person shopping? What if they're looking for something specific, such as a certain color or design? Or maybe they want to compare prices with other retailers?

If you can use AI as an assistant in these situations, you can help your customers get exactly what they need and save them time and hassle.

On a larger scale, you can use AI to recommend products and services to your customers based on their individual needs. For example, you can use AI to recommend a new car model to a customer based on their driving habits. Or, you might use AI to recommend an insurance policy for a customer based on their age, location, and risk factors.

Using AI in this way requires data that is not only actionable but also accurate. You need data about your customers' behaviors and preferences so that you can make the most informed decisions possible when recommending products or services.

Commerce.AI provides the world's biggest product and service data engine, which helps you gain this critical intelligence.

In this section, we learned about four ways to provide insights about your customers that prove to be invaluable for service teams. Once you understand your customers, it's time to zoom out and look at your locations more broadly.

Managing your locations

Now that we understand how to empower your service's front line, let's take a look at how to manage your store locations. Your store locations are a vital factor in the success of your brand, so it's vital to optimize everything you can for them. In particular, we'll look at the following:

- Optimizing your branch
- Optimizing your employees
- Optimizing your service

These three areas are interrelated, and it's important to optimize them all. Let's look at each in detail.

Optimizing your branch

People post reviews about everything, and retail locations are no exception. With the right approach, you can leverage reviews as a source of competitive intelligence to help you understand what's working well at your stores and how you can make them work even better.

You'll be able to answer questions like these: *What do customers like most about your stores? What do they criticize? Are there any common threads among their complaints?* It sounds obvious, but quite often, it's not until you look closely at an issue that you can identify how to solve it.

One powerful feature of Service AI is the ability to explore aggregated reviews. You can look at all the locations and then drill down on a subset based on attributes.

Next, we can take this approach even further by breaking down our various subsets based on external factors or categories (for example, store type: retail and restaurant). Each subset could then be scored relative to other subsets in terms of their performance so far over time: *Are some stores outperforming others? Is it clear why some are outperforming or not performing well yet?*

Once we begin to understand what differentiates the good performers from the not-so-good ones, it will help us create actionable insights such as identifying which stores could benefit most from additional training or doing something different in their marketing strategy. This will inform our decisions on where we invest our limited resources in terms of operations, personnel, and capital facilities, which are all critical things needed for any retail business.

You can also use Service AI to analyze trends across locations over time using multiple metrics, such as the number of reviews over time, the average of review scores, and more. This can also give you a line on where the sweet spot is for your locations: *Where are they performing well compared to other locations that have similar performance? Are there any outliers that don't fit this pattern? How large is their deviation from the trend, and is it growing or decreasing in magnitude over time?*

Sentiment plays a big role in understanding any store in a given branch. By analyzing sentiment across store locations, you can pinpoint areas for improvement. This isn't only relevant to physical stores but also services that vary based on location, such as a telecom service. In *Figure 10.1*, we can see a mockup of a telecom service provider using Commerce.AI to analyze sentiment by location and uncover customer intents, shown in the **Wishlist**, **Praise**, and **Warranty** sections:

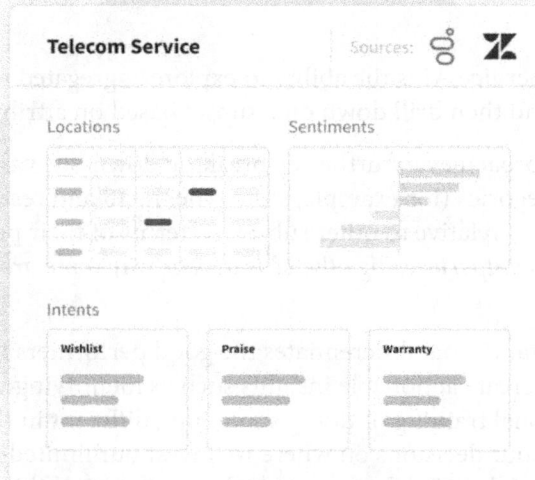

Figure 10.1 – A mockup of Commerce.AI analysis for a telecom service

Wishlists are simply features or offerings that consumers express interest in and can help inform your product or service roadmap. By aggregating a consumer's wishlist, Commerce.AI helps service teams understand what consumers want and will help guide future product development.

Praise is a great indicator of customer satisfaction, and when viewed through the lens of **sentiment analysis**, it's clear that these consumers are very satisfied. When employees interact with customers in each store location, they should be likable and helpful, and when they do something right, it's always good to celebrate!

Warranty is another key customer intent to keep track of. Segmenting customers who express interest in warranty information is a great way to understand your customers' needs and provide the right level of service.

Ultimately, these analyses come down to making wise decisions that are informed by data and analytics. By using Service AI as one of many tools for analysis, we can pull insights from several sources and help make more informed business decisions than through traditional efforts alone.

Optimizing your employees

Service innovation is a top priority for nearly every company, but one challenge is that service employees are often overlooked as an asset. The difficulty for service providers is finding the right talent and developing them into team members with the skills necessary to deliver a great customer experience.

In the following list, we'll explain how some of the world's fastest-growing companies are leveraging AI and data science to give their service teams the advantage they need to drive growth and profitability. We'll cover the following steps:

1. **Understanding where service employees can bring value**: The first step in optimizing service employees is understanding where they can bring value to your business. Often, these opportunities arise from **machine learning** or AI algorithms that provide smart recommendations for customers based on their previous interactions with your brand.

    For example, a hotel might recommend specific amenities based on a guest's travel preferences or past stay history. By using AI to create personalized interactions between customers and brands, companies can deliver exceptional experiences at lower costs than traditional marketing approaches.

    This insight holds true for all types of businesses and not just for those involved in customer services (such as hotels). Any organization can apply data science and analytics across multiple touchpoints to identify opportunities for personalized engagement that could lead to higher **lifetime value** (**LTV**) for each customer. However, when applied correctly, AI presents tremendous opportunities for product managers, designers, writers, engineers, or anyone involved in creating products or services.

2. **Sentiment analysis and customer experience analytics**: The next step in optimizing service employees with AI involves sentiment analysis and customer experience analytics. These tools can help service employees improve how they interact with customers by providing actionable insights into customer complaints, positive interactions, and other useful information for improving the customer experience.

For example, an airline could use sentiment analysis to identify the top three reasons why a customer would choose another airline over their own. By using these insights, they can improve the in-flight experience by addressing any issues that are driving customers away from their brand. This type of data science can also be used to improve retention rates and reduce **churn** (customer turnover) by proactively reaching out to customers who have started to leave your company (or stopped engaging with you altogether).

Service teams often operate in isolation from other company divisions. This is partly because traditional marketing methods don't work as well with service teams; segmentation and demographics are less relevant for them because they're focused on providing one-on-one customer interactions. However, it's also due to an unfortunate legacy of service companies being seen as subpar performers relative to their counterparts in product-focused industries.

3. **Delivering insights**: In reality, however, service organizations are just as capable of delivering profitable growth as any other business unit within any organization – and they shouldn't be overlooked. Using machine learning and AI tools like those available in Commerce.AI allows service organizations to leverage data science across multiple touchpoints so that employees can identify new opportunities for engagement and innovation at every turn.

    By applying data science across all contact points with customers, companies can better understand what motivates their users (as opposed to relying on gut instinct) and deliver exceptional experiences that create long-term value for both parties.

Optimizing your service

The services you provide to your users are at the core of your business. You don't need to be an expert in data science or machine learning to see that. The quality of your service is directly correlated to customer loyalty, and if they are satisfied, many customers will do business with you again and recommend your service to others.

That's why it is so important for service providers to leverage the growing capabilities of data science and machine learning in their businesses. In particular, sentiment analysis and customer profiling are two powerful tools that many service businesses have not fully tapped into.

Through sentiment analysis, you can understand how your users feel about the quality of your service very early in a product's life cycle and find opportunities to improve. For example, if your product is customer service, you might want to monitor social media sentiment about your team and make changes to improve the experience. Or, if your users are critical of the quality of food at a particular restaurant or hotel, you can use sentiment analysis to see if there are ways to improve that experience.

Similarly, if you are a TV show producer, you can use sentiment analysis to understand what your audience thinks of particular shows and plot a course for future programming. If your audience is tired of the same old crime shows and wants more comedy, that's an opportunity to increase viewership.

Customer profiling is another powerful data science tool that service providers can use to improve the quality of their service. With customer profiling, you can gather a wealth of information about your users – such as where they live, what they do for work, or their hobbies and interests – and use this information to provide them with customized experiences based on these preferences.

For example, if many people in one part of town watch sports highlights in real time during game day, then a sports bar might want to install high-definition TVs in that area so customers there can watch sports highlights while waiting for the game to start. This type of insight helps businesses avoid wasted investments and provide better services at a lower cost for their customers.

There's no doubt that these two forms of data science – sentiment analysis and customer profiling – will play an increasingly important role in improving the efficiency and quality of services going forward. They will become even more critical as companies look to differentiate themselves by offering enhanced customer experiences or personalized content (think **Netflix** versus **Blockbuster**).

This is why it is so important for service providers to consider how they might leverage these tools today (to improve customer experience now) to build loyal customers who will come back in the future.

In this section, we learned how to manage and optimize your store locations in terms of your branch, your employees, and your service. Each of these plays a vital role in your service. Next, let's dig deeper into using AI to improve your service offerings.

Enhancing service offerings

Enhancing your service offerings requires a holistic approach, where you look at your offerings from end to end. Your competitors are constantly improving their service offerings based on customer feedback and analysis, so it's crucial for you to do the same. In particular, you can use Service AI to enhance your service offerings in five ways:

- Identifying growth areas
- Leveraging AI for creating stronger service offerings
- Identifying opportunities to boost customer loyalty
- Finding new uses for your store
- Getting a picture of bottlenecks before they escalate

Let's explore each of these areas in detail.

Identifying growth areas

Growing revenues is hard. Businesses that have succeeded at scale have all had one thing in common: the ability to continuously generate new value and enhance existing offerings by giving customers more of what they want. In the process, these businesses have created tremendous *customer loyalty*, which has been key to their success.

Today, many companies are trying to figure out how to grow revenues in a world where there is less money around and more competition, especially as consumers look for digital alternatives to traditional brick-and-mortar shopping experiences. With e-commerce on the rise, it's no wonder why so many companies are focused on building stronger e-commerce capabilities within their organization.

But what does this mean for service teams? Now, we'll explore how we can leverage AI in order to create even stronger service offerings for our customers and potentially enhance revenue growth along the way.

Leveraging AI for creating stronger service offerings

One of the most important things that service teams can do to help their companies grow is to ensure that customers are happy. There's a lot of research showing that customer retention rates (that is, how long a customer remains loyal) are directly correlated with revenue growth. Just think about it: *If you could keep 100% of your customers, wouldn't that be great?*

But think about the implications here: *Even if you can only retain 50% of your customers, if you were able to do so by continuously providing them with more value and enhancing existing offerings, wouldn't that make sense?*

In other words, only focusing on *growth*, without looking at ways to boost *retention* rates, would be like pouring water in a leaky bucket.

What matters is whether or not your company has the ability to continue creating new value for customers over time. This means that service teams need to rethink how they go about servicing their customers in order to create new value and grow revenues at the same time.

When it comes to creating more value for your customers, you don't always have to reinvent the wheel. In fact, many companies already use AI today as a means of enhancing their existing offerings, especially on websites such as **Airbnb** and **Uber**, where people are using AI features such as score sheets and driver ratings to figure out who might be a reliable host or driver, respectively.

People are already engaging with these kinds of features on sites like these all the time; all you need is an understanding of behavioral economics (the science behind human decision-making). With this understanding, you can judge how best to deploy these tools in order for them to work optimally for your business goals at scale.

For example, let's say that you own an e-commerce site that sells home decor items such as throw pillows. Every day, thousands upon thousands of people visit your e-commerce site from around the world searching for things such as `pink throw pillows`, or even more specific terms such as `red plaid throw pillows`. Data might show that customers who buy red plaid throw pillows tend to also buy pink throw pillows, but if a customer doesn't view either, they won't buy either.

What if you added a new product detail page parameter called `color` *that would show commonly bought colors in pairs?* People who searched by color would then be more likely to buy multiple items, thereby increasing the average cart value. This is behavioral economics at work. You've used AI to make better use of your existing inventory by understanding how people actually browse through it.

The key takeaway from this example is that if you know how people browse through your site, you can use that data to inform future decisions about what products or features to include or exclude based on customer behavior. Again, using behavioral economics as a guide, you can create more value for customers by giving them what they want, when they need it most (or at least what they perceive to be the most useful). We call this **adaptive customization**.

This concept of adaptive customization has huge implications for service teams looking to create even stronger offerings and grow revenues. In fact, applying this same principle to social media platforms has been shown to increase user engagement rates significantly, which means higher retention rates and greater revenue growth over time.

Identifying opportunities to boost customer loyalty

Customer loyalty is a hot topic. Companies spend billions each year to acquire and retain customers, and the stakes are higher than ever, as there's massive competition in the service marketplace.

But with customer loyalty initiatives, which are often limited by resource constraints, it can be hard for service companies to figure out how they can give their customers more value while also boosting retention rates.

In short: *How do you create a win-win situation for both your customers and your business?*

One promising way is through AI-enabled commerce platforms such as Commerce.AI. These platforms use data science to transform the way that businesses interact with their customers through their service offerings.

For example, let's say you own a small business that offers on-demand home services to help people maintain and repair their homes. You have a great team that provides a high-quality service in a timely manner, but your revenue growth has stalled due to high customer churn.

Your sales team has been trying new ways to engage with your customers, such as creating custom events for them or offering them one-on-one coaching sessions. To add value to your existing customers, you use Commerce.AI to analyze the sentiment of your reviews on **Yelp** and Angie's List, identify what questions you could address to improve customer satisfaction, and then use that feedback to create new services for your customers.

This approach has the potential to increase customer engagement while also boosting retention rates. For example, imagine one of your customers recently complained about a broken faucet in her home. You quickly created a service offering to take care of it along with other issues she had listed on Yelp. She was thrilled with the level of personalized service and appreciative that you were able to take care of things right away.

The same principles can be applied at scale for service businesses of any size. For example, you can use Service AI to identify your top customers and then use that information to create new services for them. Or, if you have a large number of customers in certain areas, you can use Service AI to identify which customers are using certain features more than others. Then, you can create new offerings for those users based on their needs.

This kind of integration between Commerce.AI and businesses can help them strengthen existing relationships with customers by providing more value than just transactional interactions, which is why we see Commerce.AI being used by leading brands such as **Unilever**, **Netgear**, **Coca Cola**, **Suzuki**, and many others.

Finding new uses for your store

Your store may have several uses: a place to provide your service; a place to grow your community; a place to gain feedback; a place to do market research; a place to run sales promotions; a place for customers to sign up for your email list; a place to earn referral commissions; a place to sell physical products – the list goes on.

But how do you find new and better ways to serve your customers in each of these use cases and how do you track the results of your efforts?

That's where AI comes in. And because it might feel overwhelming to think about how to use AI to meet all of these diverse needs, we've broken it down into three stages that can help you start improving how you work today:

1. Finding new ways to use your store today
2. Finding new ways to use your store in the future
3. Finding out what works and identifying how to scale it

Finding new ways to use your store today

The first step is to think about how you can use your store today and then explore whether there are any new ways you can use it.

For example, if you're a service team, you might want to consider using your store as a place for customers to interact with each other. You could host a meetup or an event in your space where people can come together and learn from each other. Or maybe you have some interesting content that people would love to share, such as tips on how to be more productive at work, or how to start a business. Now is the time to start thinking about what kind of content you could create for people who want access to it outside of your newsletter or website.

You might also want to think about ways that your team could use your store as a way for them to engage with customers. For example, you could ask yourself some questions: *If you have an e-commerce site, why not host live question and answer sessions with customers? If you have a community site, why not post customer success stories?* These are just two ideas, and there are many more possibilities depending on the type of product or service that you offer. The important thing is that this exercise helps teams identify new ways they can use their platform *today*, so they don't get stuck thinking only about what they plan on doing *tomorrow*.

AI can help you find new ways to use your store by helping you identify what content is most valuable for your audience and then find the best way to deliver it.

Finding new ways to use your store in the future

Once you've started using your store in new ways, the next step is to start thinking about how you can use it in the future. This might mean starting small and building on what you're doing today, or it might mean diving into something completely new.

For example, while your stores may not currently have the budget to hire a full-time social media manager, you might be able to use your stores as a way for your team members to engage with customers on social media. You could also consider using your stores as a way for your team members to share content that's relevant to your community, such as a great new article that you think people would love, and then ask people in your community to share it with their own communities.

Ultimately, this stage is all about using AI to extract customer desires from reviews and social media posts, and then segmenting out those desires that aren't currently possible, but that can be put on a roadmap for the future. Until then, you can look for creative alternatives to meet customer needs.

Finding out what works and identifying how to scale it

The final step is to use AI to learn from your experiments so that you can scale what works. You might find it helpful to have an external team or consultant to review and validate the findings of each stage as you move forward, as this can help you stay focused on the right things.

Start small, start simple, and don't focus on the things that are most expensive for your budget (such as buying a massive piece of real estate or hiring a huge team of people) until you've figured out how to get started with something much more manageable.

This is also a good time to think about how your team will be able to track and analyze the results of these experiments over time, so they can build up an understanding of which approaches work best in which use cases, while also being able to track the success metrics behind those approaches. This helps the teams understand which experiments worked well and why they worked well, so they can repeat them in other places where there's potential for growth.

Getting a picture of bottlenecks before they escalate

Bottlenecks are the hidden factors that prevent customers from using a service to its fullest potential. These factors may stem from limited human resources, capacity, inadequate processes, or other inefficiencies.

Every business knows that improving the customer experience is key to their success. However, getting a holistic picture of all aspects of their service offerings can be difficult for many organizations. Let's explore how AI can help teams get a clearer view of their service capabilities – by isolating bottlenecks – so they can make smarter and more informed decisions about where to invest their time and resources.

Tracking your store locations, team members, and customer interactions in a centralized system can help you spot trends and identify areas for improvement. It's like having a digital *back-of-the-napkin* analysis to figure out where your business is strong and where it could use some work.

When applied to service operations, AI tools can provide real-time insights into customer satisfaction, staff performance, and location efficiency. Through machine learning, data-driven systems analyze vast amounts of data at lightning speed, crunching through reams of documents while automatically alerting you about the *issues* before they become *problems*.

In the case of service operations intelligence, AI can highlight key bottlenecks that may go unnoticed by human eyes, thereby allowing teams to proactively address these issues long before they become an impediment to the customer experience.

For example, a famous French pizza chain used Commerce.AI to identify bottlenecks at their delivery locations. Using Commerce.AI, chains can easily find bottlenecks in their business processes, such as new hires not having enough training on specific equipment.

Summary

In this chapter, you've learned how to use AI to empower your front line, manage your locations, and enhance your service offerings.

By analyzing customer data, you can extract customer affinities, purchase reasons, and challenges. This information will empower your front line to better empathize with and serve your customers. When it comes to your physical locations, you can analyze data around your branch, employees, and service to optimize processes at every level. Finally, you can enhance and even transform your service offerings by identifying growth areas, finding new uses, boosting customer loyalty, and minimizing bottlenecks.

By putting AI to work on your service data, you'll unlock new insights and opportunities for growth, and you'll further differentiate your company in the marketplace. You'll become an *AI-first company*.

In our next chapter, you'll learn how to apply AI to market intelligence to generate actionable insights that help you improve your business.

11
Delivering Insights with Market AI

The marketplace for everything is here and now. Every possible good or service that can be imagined is listed on an online marketplace, from flights to real estate, to concert tickets to old-fashioned manual labor.

It's not just something we read about in the news – it's a new reality. All around us, people are buying and selling goods and services as never before. The same forces of digitization that have democratized access to information also enable marketplaces to become ubiquitous.

It used to take companies years of investment and millions of dollars in revenue before they could even dream of creating a new, profitable product line; now, it takes weeks or months. The barrier for entry into many markets has fallen so low that competition is at an all-time high.

To succeed in this new market landscape, product teams need to be able to identify trends and patterns in their products' data that can inform them of more strategic decisions. The ability to analyze data in new ways can help product teams improve their products and offerings – and ultimately, increase revenue.

In this chapter, we'll explore how to use AI to gain actionable insights to succeed in the market. In particular, we'll cover the following topics:

- Analyzing trends and white space discovery
- Connecting market shifts to brands, products, and services
- Understanding market DNA

You'll learn how to identify new product and market opportunities, uncover and act upon market shifts, and tap into the concept of market DNA to read the voices, hearts, and minds of your customers.

Analyzing trends and white space discovery

First, let's discuss how to use AI to find megatrends in commerce through the lenses of product innovation and market intelligence.

Historically, product teams have relied upon a variety of methods to identify innovative products or features. These methods often include customer interviews, focus groups, and extensive market research. In recent years, due to the availability of additional resources and technology, product teams have been able to leverage AI-based techniques to complement or even supplant these traditional approaches.

In particular, trend analysis and forecasting have come into their own as important tools for product managers, enabling them to identify fads, up-and-coming trends, and potentially lucrative markets.

Improving product idea generation with white spaces

Idea generation starts with market research – but it doesn't end there, not by a long shot. Once a team has identified what people want through market signals, they should be able to go deeper into the data and AI, including to find so-called market *white spaces*. White spaces refer to those product segments that are not currently being served but could be if a product team can figure out how to operate in them.

In the following diagram, we can see a mockup of how white spaces are discovered by looking at data across all areas – products, services, brands, markets, consumers, and more:

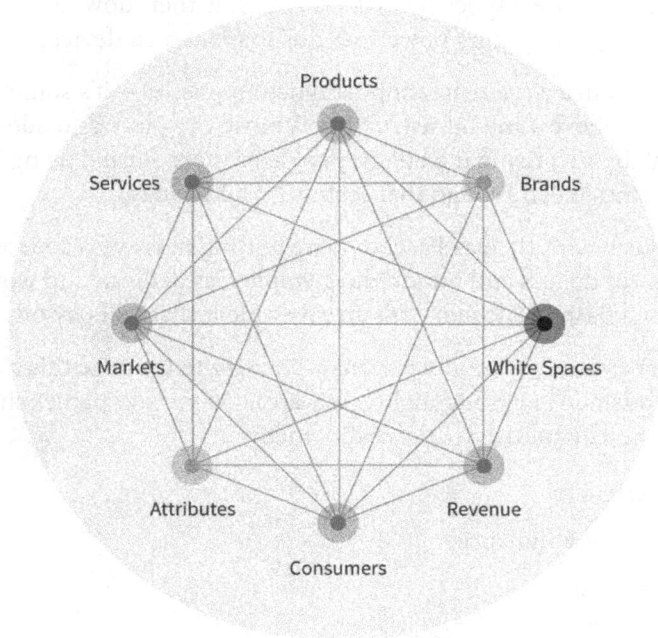

Figure 11.1 – A mockup of Commerce.AI white spaces analysis

By combining insights from data sources such as digital channels, along with qualitative information about people's actual behaviors across channels (things such as what people search for or what features they like or dislike about a particular product), product teams can develop highly predictive models that will help them create more effective products or services.

Using this concept of finding market white spaces, Commerce.AI has uncovered 10 *megatrends* across over a trillion market data points. Let's dive into these megatrends, which have important implications for product teams today.

The virtualization of everything (VE)

Virtual reality (**VR**) is the digital environment of the future for several reasons, including its immersive nature and top-of-the-market performance in terms of both hardware and software adoption. The number of VR headsets sold globally has grown over 50% in the first quarter of 2021 (`https://thejournal.com/articles/2021/07/01/virtual-reality-headsets-see-explosive-growth.aspx`), and the technology is finally starting to reach an installed base large enough to be noticed by many consumers. VR will continue to gain traction in the coming years as more people are exposed to it, driving up **average revenue per user** (**ARPU**).

These figures can only increase, especially as costs continue their downward trend while simultaneously increasing computing power per dollar spent on a device.

While VR may not seem like a mainstream phenomenon just yet – it's something you have to try before you believe – the infrastructure is mostly in place and adoption rates are rising fast. It wouldn't be surprising if a billion people or more started using VR at some point within our lifetimes (even though that seems unlikely today).

Furthermore, large businesses such as Facebook are putting massive resources behind VR, including billions of dollars and hundreds of employees dedicated to working on the technology, so it's not a flash in the pan – it's an investment that will pay off.

There are many opportunities for product teams who want to leverage this new technology for their business. Let's highlight a few areas where companies should start thinking about how they might use virtual reality today:

- Immersive experiences
- Transforming existing industries
- Branding and reputation

These three areas are prime targets for VR-based experiences and products. The opportunities are wide open but will require companies to think about entirely new forms of business models that leverage the technology's unique capabilities. We'll explore each of these areas in detail next.

Immersive experiences

Companies that offer immersive experiences will likely see the most success with VR. The main benefit of virtual reality is that it allows people to be somewhere else, whether it's a game, a movie, an exhibit, or even another country. But it goes beyond entertainment.

Companies can use VR to train their employees, put customers in touch with experts, show off products and services, and much more. As a result, companies are finding exciting ways to engage users through new channels and formats – all while offering them something they couldn't get from regular old 2D screens.

Transforming existing industries

In the past couple of years, we've seen companies use VR for aircraft design prototyping, virtual showrooms, and more. These projects demonstrate how companies are looking at older industries and finding new ways for people to engage with them using technology – in this case, virtualization.

Branding and reputation

Getting people to notice your brand is hard. That's why companies spend so much money on advertising – to reach the few people who see their ads. Virtual reality can change all of that because once integrated into a company's ecosystem, customers will want to engage with the brand. No one will want to miss out on new virtual experiences in the future; instead, everyone will want to be part of it (especially if their friends are there too).

Augmented reality

The **augmented reality (AR)** market is growing incredibly quickly as more consumers become aware of its possibilities. As with most new tech trends, AR initially targeted a niche audience – specifically, gamers looking for more immersive and realistic experiences – but as time goes on, it will continue to expand beyond this initial group.

There will be increased accessibility across multiple demographics, such as enterprise users seeking better collaboration tools, or parents who want an extra set of eyes while supervising children playing outside without fear of accidents happening.

AR will also benefit from the fact that it can be seen as an extension of VR since users are essentially living inside a digital world. The two technologies complement one another since they both provide users with new ways to interact with their surroundings – in this case, digitally augmented ones – providing them with an entirely new experience altogether.

This interaction is especially useful for those who work in industries where physical locations need to be shared with others. It's also worth noting that Facebook has recently been on an acquisition spree of AR firms, such as Scape Technologies, Daqri, and Mapillary.

Just as product teams can use VR, there are many creative uses for AR. For example, product teams can use AR to showcase new features or functions to educate users about how new features work, and even for brand engagement. Similarly, a restaurant could overlay real-time orders on top of customers' tables so that they can see how long it will take for their meals to arrive.

To give a few more examples, a real estate developer could create an AR model of a property and show it to potential buyers, while a homeowner could use AR to add virtual furniture to their homes. Consumers are even using AR to try on makeup, clothes, hairstyles, and accessories. Considering all of these possibilities, it's no wonder that companies are investing in AR to help them engage with their customers and prospects in new ways.

E-commerce

Online shopping is becoming more and more commonplace as **e-commerce** continues to experience massive growth. The Commerce.AI data engine saw 10 years' worth of e-commerce growth in just a few months in 2020, leading to tremendous e-commerce penetration all over the world.

E-commerce will also benefit from the fact that younger consumers are less attached to traditional methods of shopping or purchasing products; they see buying things online as no different than going to a mall and buying something there.

For physical stores, the biggest challenge is the so-called friction of travel. If you live in a rural area and you want to go in-store shopping, there's no other option but to get on a plane and go to a different city for your shopping trip. On the other hand, online retailers don't have this problem – all you have to do is hop on your computer or turn on your smartphone, and you can be anywhere in the world where people are willing to buy from you or send their money to you.

This shift toward online retailing has been a long time coming. A couple of decades ago, it was virtually impossible to buy anything online besides books or office supplies; today, thanks largely to e-commerce platforms such as Amazon and Alibaba, which dominate their respective markets, anyone can easily purchase almost anything with just a few clicks of their mouse.

The rise of social commerce

Social commerce has grown significantly over the past several years and shows no signs of slowing down anytime soon either, especially with the likes of Instagram and Tik Tok becoming increasingly popular as platforms for brands to connect with customers.

On the other hand, traditional media outlets such as newspapers or television ads might alienate some potential customers who aren't interested in what's being advertised at any given time (besides perhaps if it's pizza). This is, in part, why social commerce is dramatically taking over.

Currently, social media is most popular among Gen Z; however, its penetration rates have been increasing rapidly among older age groups such as Boomers and Gen Xers over the past couple of years as well. Social media has become such an ingrained part of our daily lives that it's unlikely anyone will stop spending time on these platforms – and brands that can leverage this trend will see their sales increase dramatically.

The rise of influencer marketing

Influencer marketing is another example of social commerce that has seen significant growth in the past several years. Influencer marketing spending will continue growing in parallel for the foreseeable future, especially as consumers become more comfortable with buying products based purely on someone's recommendation rather than due diligence that's been conducted by themselves.

Influencer marketing has grown alongside social commerce; both have benefited tremendously from each other as they both rely heavily on word-of-mouth recommendations among customers.

If there's one thing we've learned from the age of social media, it's that online peer pressure is often extremely powerful when it comes to influencing people and convincing them to buy something, whether they need/want it or not.

The gamification of everything

Gamification is a hot topic in terms of both commerce and marketing, with both brands and consumers alike being interested in how they can leverage the concept to increase engagement. This trend has been around for over a decade now – even Facebook was using gamification techniques since its origins – but recently the amount of attention it's been getting has increased exponentially as competition increases to grab people's limited attention spans.

The idea behind gamification is to make something that would be considered mundane or uninteresting by itself – in this case, things such as shopping experiences – into something that people look forward to doing.

For example, suppose you're selling a pair of running shoes. You could create an app for runners and turn the act of running into something interesting and enjoyable – such as competing with other people to see who can run the furthest or in the fastest time. You could also build a community around your brand and create an experience where people are encouraged to share their workout results with others, encouraging others to join them in their workouts.

This even applies to physical stores. Let's say you run a pizza restaurant. Instead of just having a pizza oven on display, you could have a game where people play to deliver pizzas to others in the community within 15 minutes. The person who gets to the top of the leaderboard wins a free pizza!

The possibilities are endless when it comes to using gamification techniques in both commerce and marketing.

The rise of the mass affluent

In financial terms, the **mass affluent** is often described as someone with liquid financial assets between $100,000 and $1,000,000.

The mass affluents are the new power brokers in global business and economy as they continue to grow in importance both economically and socially. This group of consumers is made up primarily of people who fall somewhere between upper-middle class and upper class, depending on which parameters are used to define their social status (wealth, education level, purchasing power, and so on).

Their purchasing power has grown tremendously over the past few decades as companies have been leveraging the internet to reach them and they have been able to compare price points, quality levels, and service from various consumer goods/services across many different providers. This trend has made it nearly impossible for companies in most industries to ignore these consumers – especially since they now have money to spend.

When you add all of this up over the past few years, it's no wonder that we are starting to see several trends emerge with regards to how brands interact with this group, primarily through immersive content and influencer marketing.

The rise of authenticity

With social media platforms such as Instagram, YouTube, Twitter, Discord, and so on having amassed billions of users who are interested in self-expression through self-captured content (whether its videos or still images), we're seeing a lot more authentic content being produced by users.

In addition, some of the largest brands in the world today first got their start on social media platforms before evolving into household names. This was thanks largely to their ability to connect with consumers at an authentic level.

Social platforms are not going anywhere anytime soon, so it's no surprise that many brands are trying to leverage them to create more authentic connections with their consumers. In short, consumers today are far more likely than ever before to trust and interact with brands on these platforms.

Gen Z

Generation Z (born after 1997) are the first truly digital natives and the largest generation of digital consumers. The sheer size of Gen Z is often overlooked due to their youth, but they will be entering the workforce at a time when both technology and service industries are experiencing unprecedented growth.

Gen Z has demonstrated an innate ability to understand how brands should interact with them based on their unique needs. For example, they appreciate authenticity much more than previous generations.

Demand for sustainable products

Lack of access to clean water, food, and energy in many parts of the world is pushing consumers toward buying products that have a smaller environmental impact, while also providing valuable benefits such as improved wellbeing or performance.

While this has always been an important aspect of what brands should be focusing on, Gen Z has led the way in demanding more sustainable products from brands they are interested in. These customers want to buy products that do not harm the environment but are also beneficial to them rather than compromising between both.

Companies such as Unilever have recognized this trend and created sustainable product lines with less toxic ingredients that are still effective in delivering the desired results.

To summarize these 10 megatrends, we can say that they are an inevitable consequence of the rapid evolution of many different spheres of human activity, from business to government to science. They are here to stay and will continue impacting how our lives are lived daily.

That said, understanding market shifts isn't enough – to capitalize on them, companies must be able to connect these shifts to their strategy and existing assets, as well as identify where they can innovate to better meet demand.

Connecting market shifts to brands, products, and services

Now that we've looked at ways to analyze and interpret market shifts from commerce data, let's learn how to connect market shifts to brands, products, and services. We'll look at this connection across two main areas:

- Gauge positive and negative changes in your product and product category, and respond with speed and efficiency.

- Recognize, manage, and resolve product risk areas and escalations and develop solutions proactively.

Gauging product shifts

Gauging shifts in the market is hard. How do you know what's happening and how to act on it?

It's not enough to be aware of shifts in consumer demand. You have to be able to act quickly, anticipating what those shifts will mean for your product or service offerings and how you might capitalize on them. By doing so, you can turn an opportunity into a competitive advantage that positions your company for success in the long term.

Let's look at three key ways that companies are using AI to better understand these shifting markets: identifying opportunities, gaining insights into competitors, and making changes based on data analysis.

Each of these areas is critical – and each requires AI technologies such as sentiment analysis and **natural language processing** (**NLP**). Of course, there is much more to consider than just being able to identify market shifts – but if you can leverage NLP technology as a starting point, it can unlock new possibilities for your product team.

How market shifts can help you find opportunities in the market

So, what does it mean when consumers switch from consuming water straight from their taps at home to buying pre-bottled water? Or when consumers go from buying bottled water for picnics and outdoor activities to buying it for everyday consumption? These are shifts in consumer behavior, and they're worth measuring.

How do you know what those shifts mean? And how can you leverage them to your advantage as a company – whether that means finding new ways of connecting with customers or identifying where they might be switching from your competitors?

Whether it's bottled water, electronics, or something else entirely, NLP can help us understand how consumer sentiment and interests are shifting.

This information could give us a good idea of which brands consumers plan on purchasing in the future, and this insight can then be used by product managers or marketers to create new campaigns aimed at these specific users. In essence, we would have been able to identify an opportunity in the market before anyone else.

Predicting market shifts and understanding what those shifts mean for brands and products is not just important for identifying opportunities – it's also key to creating new ones. By knowing which consumers may be interested in your product or service, you can start building out community content around that topic – all while creating a potential opportunity for your brand to attract new customers.

Gaining insights into competitors can help you create a competitive advantage

The **customer experience** (**CX**) is the new competitive advantage in e-commerce. And that makes sense – in today's world, consumers are more likely to make decisions about which brands they do business with based on how those brands treat them.

As such, companies in every industry need to pay close attention to what their competitors are doing so that they can learn from and react to what their customers want. By gaining insights into competitors through NLP technologies such as sentiment analysis, you can better understand how your brand is being perceived by consumers.

In the case of bottled water, we might be able to see a positive shift toward premium bottled water brands as people become more concerned about health and wellness. But this could also mean that there is a potential market shift away from other products in a product line, such as soda.

This information could then be used by product managers within any given company – whether it's Coca-Cola or some other consumer goods company –to determine how best to adapt their product offerings accordingly.

Making changes based on data analysis can help you adapt to shifts in the market

Data is everywhere, and more than ever, companies are using it as a way of gaining insights into their markets. As a result, it's never been easier for brands to access data on everything from consumer behaviors to competitors' sales trends. And when used effectively, this information can be a valuable tool for taking action and making informed decisions about how you should operate within any given market.

NLP technology is just one piece of the puzzle when it comes to identifying market trends and making changes based on data analysis. But no matter what industry you work in or what your role is at any given company – from marketing manager to CEO – being able to leverage NLP technologies can help you spot emerging opportunities faster than your competitors, gain deeper insights into your customer base so that you can create more meaningful connections with them, and make smarter decisions.

Recognizing product risk areas

Product risk areas are the manifestation of shifts in customer needs and behaviors that cause a product to fail or underperform. When these areas are better understood, they can be addressed proactively through design and development changes. By anticipating change, market conditions, and even competitors' strategies, product teams can develop products that meet changing customer needs.

To manage these areas effectively, product managers need to understand their customers – who they are and how they behave – so that they can anticipate what features customers might value most highly. When developing new products or services, this knowledge should inform you of how features will be designed internally and what problems/opportunities for innovation exist externally in the market.

To reduce the risk of failure, product managers should also have a strong understanding of their organization's capabilities and constraints to ensure they are building products that can be successfully delivered within their budget and timeline. They should consider how competitors are approaching similar problems, what technology will likely be used to solve those problems, and what capabilities they may lack internally but could leverage from external partners or vendors.

This kind of proactive approach allows you to identify potential risks early in the design process. It also provides you with the opportunity to build flexibility into the product roadmap by incorporating some level of experimentation, along with certain features, strategies, or business goals.

Risk mitigation should extend beyond product managers to include the entire product development team, who can work together as a cohesive unit to identify and address any potential issues that might emerge.

Product risk management with AI

The ability to proactively anticipate risk and design products that are more likely to succeed is a valuable asset for product managers, but it can be difficult for teams to manage on their own. This is where artificial intelligence can help. AI can provide a comprehensive understanding of the market by analyzing numerous data sources in real time, including social media posts and online conversations, as they pertain to the product under development.

This analysis can identify emerging trends earlier than traditional approaches through its capability to uncover patterns that may otherwise go unnoticed. The AI-generated insights that are gained from these data sources can then be continuously refined and updated over time to provide even more detailed information on the state of the market situation – all things that a proactive product manager would find useful when trying to develop an innovative product within a competitive environment.

As part of this ongoing process, new opportunities or risks can be identified and acted upon before they become full-blown problems, providing additional benefits beyond just identifying potential issues before launch or reducing failure rates once launched.

This kind of strategic intelligence could give companies an advantage in creating new value propositions that customers will want to pay for – something that was not possible when those same companies were operating without the benefit of AI technology.

To conclude this section, the use of AI technology has helped companies identify opportunities and risks in the market, as well as improving their ability to anticipate changes coming at them from all angles. Using AI in this way can help companies gain a competitive advantage by better understanding their customers and developing products that meet those needs. To succeed in a competitive market, let's explore the concept of market DNA.

Understand market DNA

Market DNA is a set of attributes and characteristics that define a market. For example, a market might be defined by the attributes of its participants – such as how many users it has or their experience level – or the attributes of what they buy and sell, such as in a subscription model.

Market DNA is essential in understanding how a product should function and what features it should have. In particular, market DNA includes key product features. For example, if you're in the smartwatch market, a key feature is battery life. So, you may want a smartwatch with long battery life to capture the attention of consumers.

The following screenshot shows the features that contribute to market DNA in the case of men's wristwatches. Innovation teams can dive into dashboard sections around **Products, Ratings, Reviews, Attributes, Consumer Wishlists,** and more. Each feature illustrates one dimension in the multi-dimensional space of market characteristics:

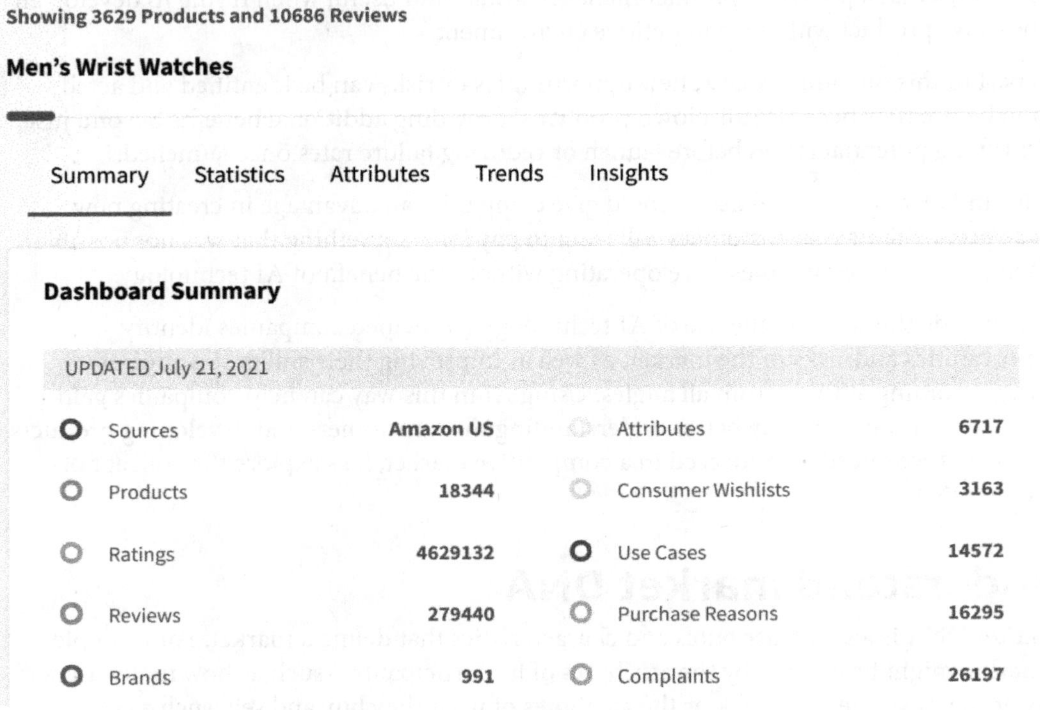

Showing 3629 Products and 10686 Reviews

Men's Wrist Watches

| Summary | Statistics | Attributes | Trends | Insights |

Dashboard Summary

UPDATED July 21, 2021

Sources	Amazon US	Attributes	6717
Products	18344	Consumer Wishlists	3163
Ratings	4629132	Use Cases	14572
Reviews	279440	Purchase Reasons	16295
Brands	991	Complaints	26197

Figure 11.2 – A screenshot of Commerce.AI's market overview for men's wristwatches

Product teams often struggle with this, as they're tasked with defining the product's purpose and building features that meet those needs. But it's rare for product teams to have data on how customers use their products – especially in early-stage companies where feature sets are still being defined. This is where Commerce.AI can help.

Commerce.AI helps product teams find patterns in market data from customers, which can uncover key product features faster than by conventional means, such as customer interviews or focus groups.

Commerce.AI also helps you understand how customers are using products, which can help product teams fine-tune their feature set and product strategy. For example, if a large percentage of users are using your product for one-off use but not regular use, this is a red flag that you might want to rethink the purpose of your product or its target user. Let's dive into these ideas more deeply by understanding market attributes, user wants and needs, finding new use contexts, and more.

Finding market DNA attributes

In this section, we will discuss how market data from Commerce.AI can help product teams find key features in the context of market DNA – including surfacing specific attributes that define the market. We'll dive into examples of key features that have been surfaced by merchants on our platform and how these insights can inform product teams' decisions.

As with many new product strategies, there's often a temptation to try and build everything for every possible user. This approach can quickly lead to feature creep and delays in shipping features as you need to add support for more use cases.

More importantly, it's not always clear which features are critical to success in a particular market. For example, is it critical for your product or service to have a long usage lifetime?

Marketers often think so, but customers may not be so attached to this. Research suggests that people generally prefer long-term products over short-term ones; however, there are exceptions – such as when the customer experience of switching is prohibitively expensive or inconvenient (for example, moving homes).

So, understanding whether customers want one-time purchases or expect their purchases to last longer might be an important insight into defining your product strategy.

Commerce data from our platform has helped merchants understand how customers respond to different pricing structures and offers on their site – which has led them to adjust their prices more efficiently while increasing revenue per customer interaction.

Through insights from Commerce.AI, these merchants have also been able to identify key features on their site that are helping drive conversion rates and increase purchase frequency among visitors.

These findings could help inform product teams about what types of features should be prioritized during development efforts, or even enable teams at other companies that are using our platform to find inspiration around what types of features they should consider building into their products/services.

Finding user wishlists and emerging needs with AI

Consumer demands are changing. People are using their smartphones more, and now they're using them non-stop for shopping. Today, we have more data than ever before about people's needs and wants – but the challenge is figuring out how to turn that information into products that people want and need.

To get there, product teams need a new way of thinking about user research. Traditional user research methods such as focus groups or one-on-one interviews often result in too much noise: lots of details that don't tell you much, if anything, about your users' underlying motivations or goals.

This can make it difficult for product teams to develop meaningful hypotheses about how to design products that meet users' needs better than existing options do. These insights will help you improve the ways your teams both start new products and evolve existing ones.

Product managers and executives have long understood that user research is essential to product development – but they've been challenged by limited access to users in the past. The best-laid plans often go awry when those plans require more human subjects than you have access to.

Today, however, advances in technology are making it possible to do what was once considered impossible: truly understand who your users are, where they are in their life cycles (compared with whom), how they think and feel about your product or service currently or potentially (compared to other options), and how you can design a product that keeps them engaged over time.

As the world's first company to design and build a platform for conducting deep market research at a global scale, we have seen firsthand how daunting this problem can seem at first glance – but also how solvable it can be once you start thinking about it in new ways.

AI and consumer-generated content

Our customers can understand their audience in more granular detail than ever before by exploring and interrogating any number of **consumer-generated content** (**CGC**) sources, such as video clips, product images, sound bites, and comments on social media posts – even contact center data, data lakes, and more.

And with our platform that connects market research teams with target audiences across all major social platforms at scale, they gain immediate insight into emerging needs and unmet desires among the people they want to influence.

In short, when you ask your customers what they want now AND what they might be interested in down the road – combined with insights gained from CGC sources such as videos and pictures posted publicly online – you're able to gain a clearer understanding of your audience's underlying needs and desires, which are often not articulated by them yet. And the more you can understand what those unmet needs or latent desires might be, the better positioned you are to create products and services that serve their long-term interests over time – whether they realize it at the time or not.

This is where we believe AI comes in: by combining CGC insights with AI technology, you're able to uncover patterns in users' behavior that you might not otherwise be able to see. For example, our customers have used our platform to identify emerging trends among specific groups of customers – such as millennials who love Android phones but hate iPhones – which can then be used for market research.

Finding new use contexts with AI

Commerce.AI uses advanced machine learning to identify the most relevant and useful use cases for a product, based on the user's prior behavior and other contextual factors. For example, if you are selling a new pair of shoes, one customer may want to wear them for exercise, while another may want to use them for weekend shopping, and yet another may just want to show them off at the club.

To successfully market and sell these shoes, you'll have to overcome the challenge of understanding how your product can be used by different users in different ways.

Product teams can then design new products with those specific contexts in mind. Product managers can also leverage this data to help prioritize features or changes within their product backlogs – prioritizing features that are likely to be more useful across multiple use cases rather than simply focusing on what is the most important for a single user or group of users.

Let's look more closely at one example: smart home products. A smart home product could be used for many different things: an energy monitoring product, a security product, a panel for your next house addition, and so on. To create the best possible experience across these use cases and to attract the most customers as a result, you'll want to design your product with some important context in mind: how will others in your user community use it?

For example, if you're creating an energy monitoring product for homes that are more expensive than average – say, $1 million or more – you may want to consider not only whether people will use your product for conservation or efficiency purposes but also whether they might be investing their money into building out other aspects of their home (such as another bedroom) – potentially increasing the demand for your product among wealthier users.

By analyzing product reviews, social media comments, and more, Commerce.AI can help you understand the context of why people are buying your product – and what other potential use cases might bring in new customers based on this information.

Summary

In this chapter, we've learned how to use AI technologies to gain strategic insights into the market, identify opportunities and risks as they emerge, anticipate shifts in consumer preferences, and reduce the risk of failure. In some cases, market shifts can be good for your business – but they can also present new challenges if you aren't ready for them.

We've explored how Commerce.AI helps firms take advantage of market shifts to create new value propositions and improve the overall health of their businesses. Using AI to gain strategic insights into the market can help companies anticipate change, identify opportunities, manage risks, and even predict how consumers will respond to products before they are launched.

Commerce.AI goes beyond simply validating hypotheses to help teams find product-market fit more quickly by identifying underserved needs in emerging markets and areas of white space within incumbent markets. Enabling product teams to better understand these dynamics enables them to invest time early on in building a viable product-market fit while mitigating the risk that comes from trying to figure out what customers want once feature sets have been defined.

In the next chapter, we'll discuss how voice surveys can be used to build an understanding of how customers think about a product or service, as well as how teams can use this information to inform innovation decisions.

12
Delivering Insights with Voice Surveys

Voice surveys are the latest big thing in customer research. This is because they are simple, fast, and easy to execute.

Unlike traditional (and expensive) text-based surveys that get poor response rates and incomplete answers, voice surveys are shown to get high response rates and in-depth responses. They can be created with just an idea and a template in just a few moments.

This is why voice surveys are quickly becoming the new standard for product innovation. And because they can be so fast and easy to deploy, companies of any size can use voice surveys to engage with users and collect feedback – improving their products more rapidly than ever before.

In this chapter, we'll learn about using Commerce.AI voice surveys to achieve the following:

- Engaging your customers with ease
- Improving your offerings
- Improving customer loyalty

With this knowledge, you'll be able to start using voice surveys to improve your product innovation workflows – fast. As a result, you'll be able to gain a competitive advantage by enabling your product teams to easily incorporate customer feedback in their workflow. Let's get started with the beginning of the customer feedback life cycle – engaging your customers.

Engaging your customers with ease

In this section, we'll explain how to use Commerce.AI voice surveys to engage your customers with ease, by creating and deploying customer surveys and questionnaires quickly using our easy-to-use templates. We'll also show you how you can implement findings in your product innovation strategies and workflows.

In particular, we'll look at how to use these seven voice survey templates:

- Product feature prioritization
- New service offering
- Post-purchase survey
- Hotel experience
- Store experience
- Post-call survey
- Pricing survey

Let's begin with an exploration of one of the most popular types of voice surveys – the product feature prioritization survey.

Product feature prioritization

Product teams are always looking for ways to make their products better. In particular, they're always seeking opportunities to improve **User Experience** (**UX**). Unfortunately, there's often a big gap between what the product team wants and what customers actually need.

This is where prioritization comes in. **Prioritization** is essentially deciding which features to build first, second, and third – or not at all! It can be challenging for product teams without any kind of formal process or structured criteria to work with. This is where **product prioritization surveys** come in handy. They're a quick and easy way for product teams to collect feedback from users while also identifying which features should get top priority next.

There are two main benefits of this approach:

- It enables users to provide valuable feedback that helps the team prioritize.

- You have a chance to observe how users respond before you jump into building something new, which could save time and money.

For example, let's say you're a consumer electronics brand and you're working on a new product – a new smartphone, perhaps. The smartphone market is notoriously competitive, and users seem to want everything: a big battery, multiple high-quality cameras, minimal bezel, waterproofing, fast charging, and so on.

You've considered these features but know that you'll need to choose just a few to deliver the best user experience. Otherwise, you'll stretch your product resources thin, and fail to deliver exceptional quality for each feature.

A product feature prioritization survey will let you know what features your users care about the most. You'll be able to see which features are the most popular, and you might even uncover some new use cases for your team to consider.

Next, let's look more closely at why you may want to run a product feature prioritization survey, how it works, and who takes part in these surveys.

What is the purpose of a product feature prioritization survey?

A survey such as this is useful because it directly engages with customers as part of your product development process. Rather than solely spending time researching competitors' offerings or coming up with ideas on your own, you can leverage existing customer insights so that your ideas are informed by what people actually want.

It also gives you data directly from real people you know more about than anyone else – namely, actual buyers who are already in the marketplace. This is invaluable because it allows you to understand exactly what people want, what they think about features, and other insights that can't be gleaned from solely researching competitors.

Companies that fail to get feedback such as this can end up making products that don't satisfy their users' needs and missing out on valuable opportunities for greater brand loyalty. For example, a large consumer electronics firm recently updated its laptop line, adding all the features that users had been requesting for several years. They've come under criticism for taking so long to deliver on these features, and for essentially building the same laptop over and over again. A product feature prioritization survey would have revealed these improvement opportunities early on.

How does a product feature prioritization survey work?

The survey template in the Commerce.AI platform is designed so that teams can quickly and easily identify which features to build next based on customer feedback. The format is simple, and the information you provide will help you prioritize the product backlog for your team.

Once users have completed your survey, it will be analyzed by AI systems that will review both the responses and preferences of users, as well as any existing market research or competitive analysis. This will give you a clear picture of what features you should work on building first, second, third, and so on.

These surveys can be sent directly to customers on platforms such as Shopify, which is how some of the largest e-commerce companies communicate directly with their customers. Since customers are the lifeblood of your business, you should always aim to get feedback from them. If you don't, you run the risk of releasing a new feature that no one wants.

Who takes part in a product feature prioritization survey?

You can send the survey to customers who have already used your product or service, or you can send it to potential customers. You can also use the survey to solicit input from employees within your organization, such as designers and developers.

When soliciting input from your team, you should offer them a chance to provide feedback on a specific feature that they're currently working on or interested in building.

New service offering

When planning to offer a new service or product, it's critical to understand the needs of your target customers, as well as the risks and challenges they face. A structured approach to customer research helps teams develop a foundation for their offering, find early adopters, and gain insights that can inform the future development of their services.

One way to understand your potential customers is by conducting customer interviews, which are an invaluable tool for gaining insights into a user's motivations and behaviors. However, conducting individual interviews can be costly and time-consuming. In addition, qualitative research methods such as observation and analysis often yield limited results, due to their low capacity for generating actionable insights that can inform the service innovation life cycle.

To overcome these challenges, many organizations now turn to voice surveys as a primary method for understanding customer needs and preferences. Voice surveys allow you to reach large audiences at a minimal cost.

Let's look at some specific examples. Suppose you operate a restaurant chain and want to understand how customers prefer to order their food. You could conduct individual interviews, but this would be costly and time-consuming.

Alternatively, you could send out a short voice survey that would allow you to reach many more customers at a lower cost and take as little as a few minutes to complete. You could also use the results of your survey to guide the development of new features that may improve customer experience and increase revenue.

Another example is a retail store that wants to understand the items that customers are most interested in purchasing via a voice survey. The store could then use this information to develop new product lines and promote those items in its checkout aisles.

This would help the store increase sales while also providing customers with additional value. As you can see, voice surveys provide service teams with a powerful tool for developing products and services that align with customer needs while also increasing revenue.

Why voice surveys are crucial for new service offerings

The service industry is facing headwinds from a variety of angles. Some of the biggest challenges service teams face include the following:

- The rise of e-commerce
- Demand for instant gratification
- The COVID-19 pandemic
- Increasing competition

The rise of e-commerce has disrupted not just product firms but even many traditional service industries, such as airlines, hotels, and restaurants.

This is due in large part to the fact that people are switching away from in-person interactions with service providers. And given how much time people spend online nowadays, this phenomenon has only accelerated.

The rise of e-commerce has also changed the way customers interact with brands in other ways – for example, when it comes to customer support. Customers now expect an instant response to their questions and concerns, or they'll simply look somewhere else for help or advice.

Additionally, today's customers expect instant gratification. It can be frustrating for an airline passenger to have to wait on hold for over 30 minutes just to speak with a customer service representative. Or it can be frustrating for a hotel guest who has to call the front desk three times before anyone shows up, only to have their room not yet made up when they finally get there.

These challenges, along with customer frustrations, have been compounded amid the COVID-19 pandemic, where customers are more concerned about cleanliness, safety, and hygiene than ever before. The need to create a customer experience that addresses these concerns is putting unprecedented pressure on the service industry.

And it's also creating opportunities for new business models and revenue streams – if service providers can figure out how to meet these needs in innovative ways. Voice surveys are a powerful tool for doing just that.

Finally, competition is an ever-present risk for service providers. To remain relevant in the minds of customers, organizations have to stay ahead of their competition by delivering high-quality experiences at a fair price. This often requires continuous investment in new products and services, which can be difficult when resources are already stretched thin.

Voice surveys give teams an opportunity to engage directly with customers on topics that are important to them – and this engagement can provide service teams with valuable insights into what's important to their customers while also identifying pain points that they need to address if they want to grow their businesses.

Post-purchase survey

The moment after a customer makes a purchase is critical. You're now on the customer's home turf, where they have the power to tell you what they like and don't like about your product or service.

To get a better sense of customer sentiment, companies use exit interviews after a purchase is made. Typically, these are in-person surveys that address roughly six main themes:

- How satisfied the customer was with their experience
- Whether they would do business with this company again
- How likely they are to recommend this company's products or services to others

- What they liked and disliked about the product or service

- What could be improved

- If anything, any other comments that stood out to them

Exit interview surveys typically take around 10 minutes and ask individuals simple questions about their experience. But it's no easy feat getting most customers to participate.

A common criticism of exit interviews is that they are self-selective, meaning that only customers who are enthusiastic about leaving a review are likely to take the time out of their day to complete the survey. But even if you get 1% participation rates, those numbers still represent a tiny fraction of your total customer base.

While those customers might be more vocal than others, it's also possible that they are less representative of the masses – and so it may not be enough for you to draw meaningful conclusions from just their opinions.

Commerce.AI has developed an online tool in which teams can conduct automated voice surveys with their customers after a purchase is made. Customers have become used to speaking freely with technology, strengthened by the rise of voice assistants such as Siri and Alexa. We've found that customers prefer these fast, effortless ways to provide feedback over the more tedious text-based alternatives.

This creates the potential for exponentially higher response rates than using traditional exit interview methods, as well as providing more detailed and actionable insights into what drives customer satisfaction and what areas need improvement.

With this powerful tool at your disposal, you can finally hone in on specific problems or concerns that customers have about your product or service so that you can begin addressing them right away – all without any overhead or burden on your team.

For example, suppose you're selling a new line of skincare products. You can set up a survey to ask customers about the efficacy of your and other products and what they liked and didn't like about using them. This enables you to understand whether you should be changing the formula or adding ingredients, or maybe even reducing the price if there is excess inventory.

In *Figure 12.1*, we can see an example of a Commerce.AI voice survey question, asking the customer, *How was your experience with this product?*. In the background, complex techniques like natural language understanding extract insights from the customer's responses, while the customer sees a simple, intuitive interface:

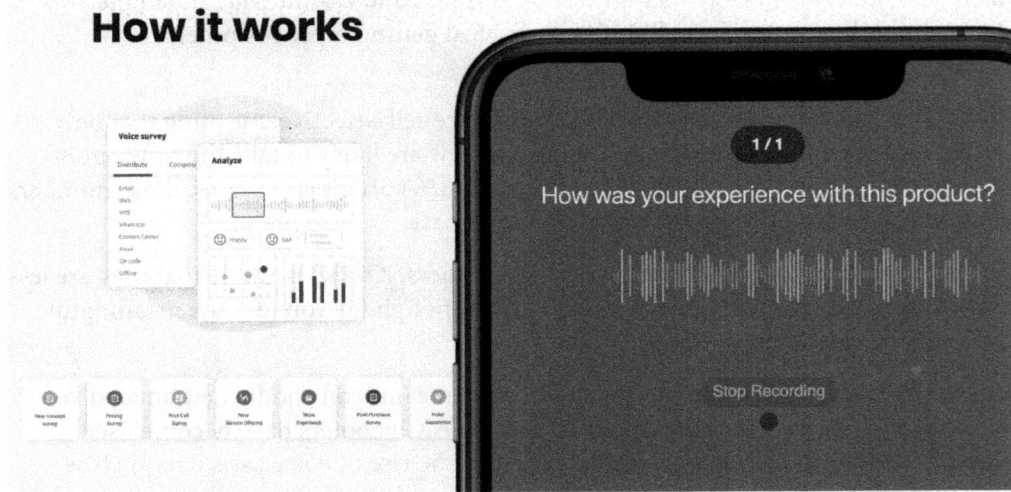

Figure 12.1 – A mock-up of a Commerce.AI voice survey

More specifically, you can use a post-purchase survey to achieve the following:

- Understand how satisfied customers are with your product (or service) and how much they liked it in comparison to other similar offerings out there.

- Collect feedback on customer service so that you can improve your support process.

- Determine whether there are any patterns associated with customers who buy certain SKUs or variations of your product, and/or, if so, learn from these patterns to help increase sales volume.

In short, voice surveys are a powerful tool that can help your business address customer issues, improve product quality, and increase sales.

Hotel experience survey

Hotels have a unique set of challenges and opportunities. They are an epicenter for social interaction, professional networking, and business travel. But hotel clientele can be notoriously difficult to engage since they have high expectations for services and amenities.

On the other hand, hotels are in a fantastic position to improve their guests' experience through digital innovation. With the right tools and strategies, digital transformation can help hotels offer unique value propositions that differentiate them from competitors.

For example, hotel brands can gain specific insights into their guests' preferences and behaviors through a hotel experience survey. Here are three ways hotels can gain meaningful insights from the survey.

Using data to improve existing products and services

The survey results can be used to inform product development and service improvements for in-room technology, food and beverage, and overall hotel experience.

For example, a hotel might want to know which amenities guests love most about its property so it can replicate those experiences whenever possible in the future. By using this type of data, brands can ensure they are meeting their guests' needs in new and creative ways while also increasing customer loyalty over time.

Suppose a hotel wants to know which features guests enjoy most about their fitness center. It could use the results of the survey to guide its investment decisions in this space, such as by choosing equipment that will be effective at engaging users and ensuring they have a positive experience. For example, the hotel might decide to invest in treadmills rather than elliptical machines if results show guests prefer running machines more than walking ones.

It could even be something as simple as making small aesthetic changes to the fitness center, such as adding art or better lighting, that will help guests enjoy their experience even more. Whatever the improvement may be, voice surveys will highlight what customers care about, so you can allocate your resources based on what will have a real impact on customer satisfaction.

Identifying opportunities for new revenue streams

Hotels can use the survey results to identify new revenue streams. For example, a hotel might discover that guests enjoy using the hotel's business center, even though they don't plan to meet any of their associates or attend a conference. This could mean the hotel should invest in additional meeting rooms to capitalize on this behavior and turn it into a revenue stream.

Perhaps the business center could be upgraded to a modern, high-end space that matches the rest of the hotel and offers guests Wi-Fi, comfortable seating, and other amenities. This might not only help generate additional revenue for the hotel, but it could also make business travelers more loyal to the brand — and, therefore, more likely to stay at that brand's properties in the future.

By doing so, you'll boost the bottom line while increasing customer satisfaction at the same time.

Ensuring consistency across properties

The survey results can also be used to inform future product rollouts and even franchise agreements if applicable. For example, a hotel brand might want to know which amenities guests love most about their property so they can replicate those experiences whenever possible in other properties owned by the same brand or even other brands.

This is known as *parallel innovation*, and it's an effective way of spreading innovation across multiple brands while also ensuring they have similar offerings. Using survey data throughout the life cycle of a project ensures all teams are aligned on how to best achieve goals and deliver an innovative product or service to customers.

Store experience survey

When consumers shop online or at physical stores, the experience they have can make or break a purchase. More than half of online consumers say that poor customer service at a retailer is likely to discourage them from making a repeat purchase.

Because the goal of any store is to drive sales, it's critical for retailers and brands to understand what factors influence customers' likelihood to buy, how their shopping experience could be improved, and whether they are satisfied with the store or product itself.

Let's dive into a specific example. Suppose you operate a pizza chain and have just launched a new store design. You want to understand whether customers like the new look and feel of the store. You can use the Commerce.AI store experience survey to understand what aspects of your new design are most likely to influence customers' likelihood to buy, how they feel about your new look, and whether they are satisfied with their overall experience. These insights can help you prioritize your store redesign efforts and ensure that you are investing in the right elements.

You can also use the survey to understand how customers feel about specific aspects of your new design, such as lighting or music. For example, if customers don't like the sound of a cash register ringing when they enter a pizza shop, you may want to consider a *noise-free* register system.

Another way to leverage this data is to identify patterns and trends that could indicate areas for improvement. For example, if customers consistently report being dissatisfied with their overall experience at your pizza shop, it may be time to invest in an overhaul of your ordering process.

A store's overall customer experience has the potential to drive future purchases, which is why retailers and brands are increasingly turning to data-driven approaches to understand how their customers navigate through their physical spaces.

Retailers can use analytics platforms such as the Commerce.AI store experience survey to survey customers in-store or online about how they feel about the store itself, as well as specific elements of the shopping experience. By gaining a better understanding of what factors influence customer behavior and satisfaction, retailers can make informed decisions about investments that could improve customer loyalty, engagement, and ultimately sales.

Post-call survey

Contact centers have long been a staple of providing customer support, and the rise of chatbots and voice bots has only reinforced that trend. Today, many consumers use contact centers to interact with businesses — and those interactions often lead to sales.

When a customer calls a business — whether they're reporting an issue with an order or seeking help with their purchase — businesses would be wise to collect that information for insights later on. You can do this by conducting a quick post-call survey to understand what (if anything) went wrong and how to improve the customer's service experiences moving forward. This helps organizations prevent costly mistakes from happening by proactively addressing issues before they arise.

A post-call survey can also provide insight into the quality of the overall customer experience, which can influence future revenue opportunities (and losses). For example, if you identify that your call agents aren't properly following up on issues, it may be time to invest in automated reporting capabilities to ensure your agents are doing everything possible to follow up on issues — especially if this is impacting your bottom line.

In addition, these insights can help you prioritize efforts as you develop new products or services for your customers. For example, if you learn that certain features are significantly more important to customers than others based on their responses in a post-call survey, this can help you prioritize your product roadmap.

Like the store experience survey, a post-call survey can also provide insights into how customers feel about specific elements of your service delivery, such as responsiveness or agent knowledge. By gaining a better understanding of the factors that influence customer behavior and satisfaction, you'll be able to make informed decisions about investments that could improve customer loyalty and engagement.

Pricing survey

The prices of your products and services are critical to your business. They can create a barrier of entry for new customers, influence consumers' willingness to pay, and signal value to existing customers.

If you sell a product or service that is priced too low, it will be difficult to capture enough revenue to be profitable. If you sell a product or service that is priced too high, it will likely result in lost revenue and profitability. Pricing is really important!

That said, pricing decisions can be complex and challenging. Many companies struggle with pricing because they don't have the right tools or processes in place for setting prices, making strategic adjustments, or anticipating potential price objections from their customers.

Product innovation teams have a huge opportunity to improve their pricing performance by leveraging the insights from the Commerce.AI pricing survey. The survey is designed to help you understand your pricing strategy and identify opportunities for improvement.

We've provided a template that you can use to quickly design your own pricing survey, or you can use our predesigned example to get started. Either way, we hope that this template will help you engage with your customers on pricing questions and inspire new ways of thinking about how and where to set prices in your products and services.

Why price is key

Why do you need to know your pricing strategy? Many companies have a pricing strategy, but few have a robust conversation around it. Pricing is ultimately about creating an equilibrium between your cost of goods and the value that customers place on those goods and services.

Being able to set prices for your offerings helps you determine how much revenue to generate from each customer segment. It also helps you identify the most profitable products and services that can be leveraged across multiple business units or lines of business. Finally, understanding your pricing strategy can help you make better strategic decisions about investing in new products and services.

Now that we've learned how businesses can deploy voice surveys in a variety of areas, let's explore how to implement your findings to improve your offerings.

Improving your offerings

Your existing products and services are, of course, the foundation of your business. But, if you're like most companies, your offerings are likely to have some gaps. Until now, though, there was no good way for you to know about these gaps.

That's why we created voice surveys. With voice surveys for market research and analysis, you can derive insights into your existing products and services, and help formulate new ones. Let's dive into how to use voice surveys to gain insights into your existing products and customers, and how to leverage those insights to improve your offerings.

Deriving insights into your existing products

Creating and launching a product is only part of that product's life cycle. It's also crucial to understand what makes it succeed or fail.

But in today's fast-moving world, even the most seasoned executives may not have all the information they need to make critical business decisions. So, when you can't find enough data in your existing market research, you need voice surveys.

By answering questions about your product through a survey, you not only get more data but also gain insights into what customers are really thinking. You do this by asking users to vocalize their thoughts and feelings, which is far more engaging than having them respond with one-word answers on a text form.

This insight will help inform better future product development and improvement efforts – and ultimately boost the chances of success for each new offering. Let's explore how you might use this approach to improve both your current products and services, as well as your future offerings.

How to leverage survey feedback to understand your customers

Before you embark on any kind of customer journey mapping project, it's important that you understand what motivates your target customers so that you can devise strategies that will help you increase conversion rates and reduce friction points along the way.

This is where customer surveys come in handy because they allow you to pinpoint pain points that are unique to each segment of your target audience(s). For example, if you have a **Software as a Service (SaaS)** product that focuses on small businesses, then your customers may be most concerned with costs and pricing. In this case, you'll want to incorporate questions about pricing into your survey so that you can understand what's important to your target customers.

You should also ask customers to rate their satisfaction with your product by answering questions such as the following:

- How likely are you to recommend our product to a friend?
- What's the most common reason for switching from our competitors?
- What features do you wish we offered?

By doing this, not only can you understand what motivates those who use your product today but also those who might become future users. This kind of data will help you improve future products and offerings.

How to act upon those insights

Once you've identified where your customers are feeling pain points or experiencing friction, it's time to move forward with a solution. But first, it's important that you understand the difference between *growing pains* and fundamental problems with your product.

If one of your survey respondents said they had a problem with something such as late deliveries or poor customer service but they still planned on buying from your company, this is likely a growing pain – an issue that will be resolved once the company scales to meet demand.

If however, respondents said they were dissatisfied with their last purchase experience and were planning on never purchasing from your company again, this is a fundamental problem – one that needs to be solved if your company wants to succeed in its current market position.

The good news for product innovation teams is that most growing pains are avoidable; they're often caused by a lack of understanding around what customers actually want out of an experience.

By really getting inside the minds of your customers through surveys and listening closely to what they say they need, you'll be able to deliver on the promise of your business. As such, it's crucial to ensure that insights from surveys are shared within your company as a means of understanding your customers and building better products.

This will help everyone from product managers to data scientists to gain a better understanding of what motivates your customers – and why – so that you can build with confidence, knowing that you're working with the most valuable asset in any business – your customers.

Coming up with new product ideas

Beyond deriving insights into your existing product line, you'll also want to consider what new products and services you can bring to the market. That said, new product ideas are hard. And the more successful you are, the harder it gets, given the tendency for companies to become complacent and avoid risk.

Most products are developed incrementally, based on customer feedback. The process can be iterated over and over again, with each iteration improving upon what came before. But what if you're a company that has historically focused on solving one specific problem? What if you're trying to come up with something totally new? How do you even start? These questions are often what keep product teams up at night.

With Commerce.AI voice surveys, you can gain insights into your customers' pain points and what they expect from different types of products.

This information will help you better understand the needs of your target customers so that you can identify which ones are most likely to buy your new offerings – as well as inform your next steps, including product development. Let's dig into a specific example, with anonymized company names.

Example – a tech start-up developing a new product

A tech start-up called SportFlix has built an impressive *following* on Instagram and other social media platforms, where its fans share photos and videos of their favorite sports moments using the company's app. The app also helps fans stay engaged with their favorite athletes and teams – and it can even notify users about important events, such as injuries and changes to lineups.

SportFlix is now looking to expand into other areas that involve watching sports, such as playing fantasy sports or viewing scores from games that aren't broadcast live. This will require the company to create new products for different kinds of sports lovers, who may have different needs than those who use its existing product.

The team at SportFlix wants to understand what types of products are most appealing to different audiences so that it can develop offerings that better meet the needs of these groups.

To do this, they'll need to answer some questions through a voice survey so that they can come up with a list of potential new products. Here's what the team might ask:

- What sports do you watch?
- What are your favorite teams and players?
- What do you enjoy doing during a game or when watching a sport?

- How often do you participate in fantasy sports leagues?
- What would motivate you to use our product instead of other apps?
- Do you have any pain points that we can solve with our existing product, or with future iterations of our product?

By asking these questions, the SportFlix team will be able to gain an understanding of its target audience's interests and preferences – and then use that information to inform future product development efforts. This will result in more successful ventures for the company, as well as higher customer satisfaction and retention down the line.

For example, let's suppose that Commerce.AI's insights reveal, through an analysis of survey responses, that SportFlix fans wish an app existed that they could use to socialize with other fans during a game, as well as check on the latest scores. The team will then be able to use this information to help them identify which features they should build into their product next, or even whether they should build an entirely new app.

While improving your products is vital to staying competitive, customer loyalty can be just as important. With that said, let's take a look at how to use voice survey insights to increase customer loyalty.

Improving customer loyalty

Customer loyalty is important because it helps companies grow and retain customers over time. In turn, growing and retaining a customer base will help a business to succeed – and a company that fails to meet the needs of its customers will eventually suffer from a decline in loyalty.

Indeed, poor customer loyalty results in high rates of churn and decline, which can lead to significant financial losses and a company's eventual demise.

But customer loyalty isn't just about keeping customers; it's also about attracting new customers. To do this, companies must be able to offer products and services that are truly valuable for their target customers – and that means becoming an expert on what those customers want.

With voice surveys, you can obtain feedback directly from consumers, wherever they are. With increased response rates and engagement, you can better understand your customers and their needs, which will allow you to create offerings that address these needs and improve both customer loyalty and customer acquisition.

The key is to make the process as easy as possible for both consumers and company employees. By using Commerce.AI voice surveys, you can gain insights into your target customers' needs in order to help you better serve their needs, while also gaining valuable information about them that can be used to attract new customers down the line.

What drives customer loyalty?

Increasing customer loyalty is essential to sustaining high levels of revenue growth. Let's look at the four main pillars of customer loyalty:

- Generating brand affinity and positive emotions toward the brand
- Increasing purchase frequency and reducing customer churn
- Creating brand advocacy through **Word-of-Mouth (WOM) marketing**
- Increasing the revenue generated from existing customers

Let's explore each of these pillars to better understand how to use insights from voice surveys, and beyond, to drive customer loyalty.

Generating brand affinity and positive emotions toward the brand

The actions consumers take when interacting with your company's offerings can have a profound effect on their perception of your brand and how they view it in relation to other brands.

If you're operating a hospitality business like an inn, for example, then having happy customers who enjoy their stay will help you build brand affinity and drive retention down the line. The quality of their stay is impacted by a number of factors, including the responsiveness and courtesy of your front desk staff, the cleanliness of their rooms, and the quality of the food and beverages they consume.

Similarly, if you're a restaurant, customers who enjoy their meal experience will be more likely to recommend your business to others – and they'll also be more likely to return in the future.

Brand affinity and positive emotions help consumers make buying decisions that result in increased brand loyalty and purchase intent. This is why it's crucial for companies to listen carefully to their customers; only then can you understand what they really think about your company and how they view it compared to other brands in your industry.

Voice surveys are a powerful tool to find out exactly what your customers want, and they can also help you build brand affinity and positively influence customer loyalty.

Increasing purchase frequency and reducing customer churn

One of the best ways for a company to increase revenue is to grow its customer base –and this starts with increasing customer loyalty through increasing purchase frequency and reducing customer churn.

Customer loyalty is highest when customers perceive that their needs are being met and they have a positive experience with your company's products and services. To achieve this, you need to make sure that employees understand what truly qualifies as *meeting a customer's needs*, which includes everything from product availability to price point and payment options.

It also means uncovering customer sticking points and coming up with solutions, which is where voice surveys come in. By asking your customers about their pain points, you can identify areas where they are having difficulty – and then come up with ways to solve those problems. This will help to reduce customer churn and increase customer retention because your customers will be more likely to recommend your business to others.

Creating brand advocacy through WOM

If you want consumers to tell their friends about your brand, then it stands to reason that those friends will trust and respect their consumer friend more than they would someone who isn't connected with them – which means good old-fashioned *WOM marketing*.

Asking questions about what different people value most about certain brands or how often they interact with each brand will help companies determine whether they're successfully creating brand affinity among their target audience via WOM. If not, then there may be room for improvement on one of the pillars of customer loyalty.

Research indicates that customers who are referred through WOM are significantly more loyal customers than those who are referred through other means, such as advertising or social media. This is because customers trust and respect their friends –and those friends will be inclined to tell others about great experiences they've had with your company.

Increasing the revenue generated from existing customers

It goes without saying that improving customer retention is a great way to increase revenue in the long term – but it's also essential in growing your customer base.

To help increase revenue from existing customers, you'll want to make sure that they're satisfied with your company's offerings and services, as well as any additional products or services you might provide them down the line.

Providing excellent customer service will go a long way toward helping consumers feel like they can trust you and your brand – and once they do, it's easier for them to share their positive feelings about your brand with others.

From the customer's perspective, this means that more of their problems are being solved, which can lead to higher levels of customer loyalty. Ultimately, voice surveys are a powerful tool to improve customer loyalty across each of its pillars.

Summary

In this chapter, we've learned about how voice surveys can help companies gain insights into customers' needs, wants, and concerns. By answering questions about their pain points or what they value most about certain brands or products, consumers are able to provide you with valuable information that you can use to develop new offerings or improve existing ones.

By using Commerce.AI's voice surveys, businesses of all types can gain actionable insights into their target markets – and use this information to better understand their customers so that they can continue growing and improving for years to come.

Whether you operate a hotel chain, a number of stores, a software company, or work in an entirely different industry, voice surveys are a powerful way to gain insights into your market and better serve your customers.

As you've learned, the Commerce.AI platform is a powerful mix of machine learning and big data that can be applied to the innovation needs of product and service teams. The platform provides an integrated suite of analytics and AI capabilities that can be applied to better understand customer needs and preferences, predict future trends, and optimize product and service portfolios. We encourage you to use what you've learned to innovate and to consider how Commerce.AI can help your organization.

Packt>

Other Books You May Enjoy

If you enjoyed this book, you may be interested in these other books by Packt:

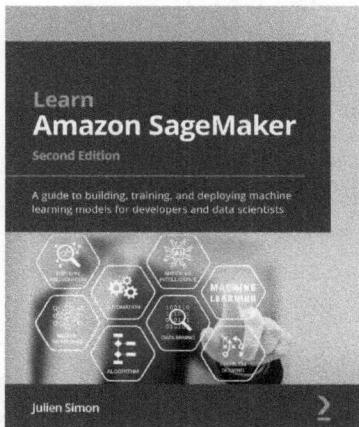

Learn Amazon SageMaker - Second Edition

Julien Simon

ISBN: 9781801817950

- Become well-versed with data annotation and preparation techniques
- Use AutoML features to build and train machine learning models with AutoPilot
- Create models using built-in algorithms and frameworks and your own code
- Train computer vision and natural language processing (NLP) models using real-world examples
- Cover training techniques for scaling, model optimization, model debugging, and cost optimization
- Automate deployment tasks in a variety of configurations using SDK and several automation tools

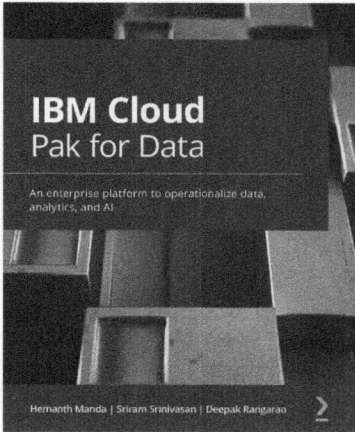

IBM Cloud Pak for Data

Hemanth Manda, Sriram Srinivasan, Deepak Rangarao

ISBN: 9781800562127

- Understand the importance of digital transformations and the role of data and AI platforms

- Get to grips with data architecture and its relevance in driving AI adoption using IBM's AI Ladder

- Understand Cloud Pak for Data, its value proposition, capabilities, and unique differentiators

- Delve into the pricing, packaging, key use cases, and competitors of Cloud Pak for Data

- Use the Cloud Pak for Data ecosystem with premium IBM and third-party services

- Discover IBM's vibrant ecosystem of proprietary, open-source, and third-party offerings from over 35 ISVs

Packt is searching for authors like you

If you're interested in becoming an author for Packt, please visit `authors.packtpub.com` and apply today. We have worked with thousands of developers and tech professionals, just like you, to help them share their insight with the global tech community. You can make a general application, apply for a specific hot topic that we are recruiting an author for, or submit your own idea.

Share Your Thoughts

Now you've finished *AI-Powered Commerce*, we'd love to hear your thoughts! Scan the QR code below to go straight to the Amazon review page for this book and share your feedback or leave a review on the site that you purchased it from.

`https://packt.link/r/180324898X`

Your review is important to us and the tech community and will help us make sure we're delivering excellent quality content.

Index

T

U

www.ingramcontent.com/pod-product-compliance
Lightning Source LLC
Chambersburg PA
CBHW061401210326
41598CB00035B/6064